CAMBRIDGE COUNTY GEOGRAPHIES

General Editor: F. H. H. GUILLEMARD, M.A., M.D.

MONMOUTHSHIRE

CAMBRIDGE UNIVERSITY PRESS
Cambridge, New York, Melbourne, Madrid, Cape Town,
Singapore, São Paulo, Delhi, Mexico City

Cambridge University Press
The Edinburgh Building, Cambridge CB2 8RU, UK

Published in the United States of America by Cambridge University Press, New York

www.cambridge.org
Information on this title: www.cambridge.org/9781107674325

First published 1911
First paperback edition 2013

A catalogue record for this publication is available from the British Library

ISBN 978-1-107-67432-5 Paperback

Cambridge County Geographies

MONMOUTHSHIRE

by

HERBERT A. EVANS, M.A.

With Maps, Diagrams and Illustrations

Cambridge :
at the University Press
1911

Gwyllt Walia ydwyt tithau, Mynwy gu!
Dy enw'n unig a newidiaist ti.—Islwyn.

[Wild Wales art thou still, dear Monmouth!
Thy name only hast thou changed.]

CONTENTS

CONTENTS

ILLUSTRATIONS

MAPS

The Illustrations on pp. 3, 6, 9, 11, 14, 16, 17, 18, 25, 34, 42, 98, 100, 104, 106, 109, 115, 116, 118, 119, 123, 125, 128, 137, 139, 140, 142, 150, 164, 166, 168, 169, 171, 172 and 175 are from photographs supplied by Messrs Frith and Co.; those on pp. 22, 26, 56, 60, 64, 66, 72, 78, 102, 122, 131, 133, 144 and 161 from photographs by Messrs Valentine and Sons; those on pp. 63, 74, 75, 76 and 77 are reproduced by kind permission of the proprietors of the Great Western Railway Magazine; those on pp. 80, 93 and 94 are from photographs by Mr H. L. Jones; those on pp. 129 and 174 are from photographs by Mr W. E. Call; that on p. 20 is from a photograph by Mr Ballard and the portrait on p. 149 is reproduced from a photograph by Mr Emery Walker.

For the Section on the Geology of Monmouthshire, pp. 27—35, the author desires to thank Mr James G. Wood, M.A., F.G.S., who kindly contributed a similar section on the Geology of Gloucestershire in a previous volume of this Series.

1. County and Shire.

From the time of Edward I to that of Henry VIII, the only part of the country now called Wales which was directly governed by the English king was, roughly speaking, that which now forms the modern counties of Flint, Anglesey, Carnarvon, Merioneth, Cardigan, and Carmarthen. The rest consisted of the Marches, a term comprising numerous districts, each of which was governed by its Lord. These "Lords Marcher," as they were called, had been permitted to conquer their territories from the Welsh from Norman times onward. Each ruled his own domains like a petty king, administered justice in his own courts, and owed allegiance to no one but the king of England. Edward IV was the first to interfere with the power of the Lords Marcher, and under Henry VII a Court of the Council of Wales and the Marches was fixed at Ludlow. To this Court the jurisdiction of the Marcher Courts was gradually subjected, until in 1535 Henry VIII finally abolished the authority of the Lords Marcher and distributed their Lordships among five new counties[1], though some of them were added to the older Welsh or English shires. In this way

[1] Monmouth, Brecon, Radnor, Montgomery, and Denbigh.

part of the Lordship of Striguil, viz. the parishes of Tidenham, Lancaut, and Woolaston, being east of the Wye, were added to Gloucestershire.

The twenty-four lordships which occupied the ancient Gwent as well as Gwynllwg[1] (Wenllwch), the county between the Rhymney and the Usk, were then taken to form the county of Monmouth, which was declared for the purposes of jurisdiction to be an English county, and placed under the courts of Chancery and Exchequer at Westminster. About the same time a court entitled The King's Great Sessions in Wales was established for the rest of Wales and the county of Chester. This court lasted down to 1830, when the North and South Wales Circuits took its place, but Monmouthshire had already been placed on the Oxford Circuit by Charles II. The result was that from a legal point of view the county became, and is still considered, an English one, but in all other respects it remained Welsh, and to this day in statutes affecting Wales, the words "and Monmouthshire" have to be added. This awkward addition would have been saved if the framers of the Act of Union had accepted the facts and boldly made their Welsh counties thirteen instead of standing out for the round dozen.

Thus we see that the Mercian shires had been in existence some five or six hundred years, and the counties of Wessex for about a thousand years, before Monmouthshire was heard of. Like the other new counties it was divided into hundreds on the English pattern, and

[1] That is, the district of Gwynllyw, just as Morganwg is the district of Morgan.

Monmouth

I—2

Gwent and Gwynllwg (Wenllwch) with their ancient divisions into cantrefi and cwmmwds were forgotten.

The old English word was "shire" (a part *shorn* off), the division of the country administered by a shire-reeve or sheriff; but about the beginning of the fifteenth century the Anglo-French word "county" (*comté*), the territory of a comte or count, came to be used as its equivalent. Hence we may now speak indifferently of " Monmouthshire " or "the county of Monmouth," but not of the " county of Monmouthshire." The hundreds (in some counties represented by "wapentakes," in others by " wards ") were subdivisions of the county for administrative purposes. The origin of the term is disputed, but a commonly received explanation is that it denoted a district inhabited by a hundred families. As applied to Wales it is of course an exotic.

Our county following in the wake of the Mercian " shires " has taken its name from its chief town. The word Monmouth means, of course, the town at the mouth of the Monnow, Welsh Mynwy, just as Plymouth and Exmouth are the towns at the mouths of the Plym and the Exe. The Welsh name of the town is Trefynwy (the town on the Mynwy).

2. General Characteristics.

A modern Welshman, especially if he hailed from Morganwg, might object that there was something anomalous about the geographical constitution of the county of Monmouth. He might urge that the eastern extremity of the great mineral-producing mountain system should not be cut off from the rest, and should rather be included in his own county of Glamorgan. Nor would such a contention be unsupported by existing facts. The Usk, or at any rate the Afon Lwyd, seems to form a more natural boundary than the Rhymney. In our day considerations also of language as well as geographical conditions alike point to this conclusion. West of these rivers the country is mountainous and the language Welsh; east of them Welsh is rarely spoken, and the country is undulating only, or rising at most to 1000 feet. Nevertheless, at the time when the shiring of South Wales took place, one at any rate of these distinctions had less force, and the Rhymney remained as before the eastern boundary of Glamorgan, while the two ancient districts of Gwynllwg (Wenllwch) and Gwent were taken to form the new county of Monmouth—a combination which has to-day given Monmouth its place among the industrial counties of South Wales together with its industrial capital of Newport, the latter advantage one which Glamorgan, with its Cardiff and its Swansea, need not grudge us.

A hundred-and-fifty years ago this mountainous region was hardly to be distinguished from the wide stretches of hill and pasture north of the Usk, and it was only here and there that the valleys were marred by the smoke of the charcoal then used for smelting the iron ore, but the woods which had furnished the charcoal had gradually been cleared or destroyed by the numerous herds of goats which ranged through them at will, and about that time the discovery was made that pit coal answered all the

The Usk and the Sugar-loaf

purposes of charcoal. The iron industry then took a fresh start and these valleys soon became the home of a teeming population of miners and mechanics. At the present day the number of inhabitants is hardly less than that of any area of the same size throughout the South Wales coalfield.

The rest of the county is wholly occupied with grazing, agricultural, and wood land. The hills of the east and south-east are covered with heathy commons, vast woods, and modern enclosures which, under the various names of Penallt common, Trellech common, Wye's Wood common, Chepstow Park, Earlswood, and Wentwood, stretch from Monmouth to Caerleon. The mountains of the north-west, above the line of cultivation which struggles up their lower slopes, are grazed by hardy flocks of small Welsh sheep, while the undulating tracts of the centre and the rich levels of the south are covered with small farms. Lastly the Severn Sea, which forms the southern boundary of the county and receives the tidal estuaries of the Usk and Wye, constitutes our claim to be considered a maritime county.

3. Size. Shape. Boundaries.

The shape of Monmouthshire is that of an irregular parallelogram with a projection on the north comprising the valley of the Honddu. The extreme length from the house called "The Monmouth Cap," once an inn, near Grosmont, to Goldcliff on the Bristol Channel is about

30 miles, and the breadth from Bigsweir on the Wye to the Rhymney is about 23 miles. The area, excluding water, is 347,600 acres, or about 543 square miles.

A perambulation of the county boundaries would conduct us through varied scenes of hill and vale, seashore and forest. Starting on the Wye some three miles north-east of Monmouth at a point opposite to the Herefordshire hill called the Great Doward, famous for its prehistoric caves, the boundary may be followed southward in a zigzag line through the outskirts of the Forest of Dean down to Redbrook, where it again meets the Wye. In this part of our course we have traversed the depression which separates the Buckstone (915 feet) in Gloucestershire from our Monmouthshire Kymin (840 feet). From this point the Wye itself forms the boundary dividing our county from Gloucestershire. On either side of the river lofty wooded hills, interspersed below Tintern with precipitous limestone cliffs, give characteristic features to the valley, and only cease altogether less than a mile from the river's mouth.

From the mouth of the Wye the Bristol Channel forms the boundary as far as the mouth of the Rhymney, which lies nine miles south-west of Newport, and about one mile east of Cardiff. The sea wall and the other features of the coast are described in a later chapter. It may be noted here that the width of channel which at the mouth of the Wye is only two miles, is nine miles at the mouth of the Usk, while the mouth of the Rhymney is eleven miles from Weston-super-Mare, and twenty from the estuary of the Parret. The counties of Gloucester and Somerset on the opposite coast are divided by the river Avon.

The Rhymney forms the western boundary of the county from its mouth to a place called Rhyd-y-Milwyr, close to its source, where the three counties of Glamorgan, Monmouth, and Brecon meet, and the last named becomes our frontier till it marches on Herefordshire in the moun-

Valley of the Clydach, Breconshire Border

tains above Llanthony.   At Rhyd-y-Milwyr the boundary strikes across the open mountain for two miles and a half[1], and then turns south-east and east as far as Careg Maentarw, leaving Brynmawr and the Clydach to the north of it.   Here it turns northward once more, crosses

[1] This piece is a modern extension of the county at the expense of Breconshire.   The boundary formerly turned east at Rhymney Bridge, and is now resumed about half a mile west of Brynmawr.

the canal, and reaches the Usk at Aberbaiden, two miles above Abergavenny. It then ascends the Usk for two miles, when it turns to the north-east up Cwm Gwenffrwd, and crossing the western flank of the Mynydd Pen-y-fal (Sugar-loaf) arrives at Pont Newydd in the valley of the Grwyne Fawr, an affluent of the Usk. It then ascends this valley—a wild and remote glen—climbs to the summit of the Chwarel-y-Fan (2228 feet), and in another mile descends into the valley of the Honddu at a point just below Capel-y-ffin, and three miles above Llanthony. It then climbs the opposite hill to its summit (2091 feet), turns south-east, and follows the ridge for five miles, when it descends the steep slopes of Hatteral Hill into the valley of the upper Monnow.

Thus after passing through the heart of the coal district up the valley of the Rhymney, and threading a devious course up hill and down dale for some twenty miles, we find ourselves again in the basin of the Wye at the Honddu. Another ten miles brings us to its confluence with the Monnow, and this river now forms the northern boundary of the county as far as Perthir (Perth-hir, the long brake) two and a half miles above Monmouth, except for a mile at Skenfrith, where the boundary line diverges for a mile in order to include in the county that part of the parish which lies on the left bank of the river. The course of the Monnow when it first becomes the boundary is north-east, but at the Monmouth Cap, less than a mile to the south of Pontrilas station, it bends to the south-east and falls into the Wye just below Monmouth. But at Perthir, in order to

include the town of Monmouth, the boundary leaves
the northern bank of the river, and after a zigzag course
over well-wooded hills, meets the Wye two miles below
our original point of departure.

A county formed as late as 1535 would not be likely
to have many detached portions, and there are now none;
but the parish of Welsh Bicknor on the right bank of the

Mill on the Monnow near Monmouth

Wye, which had belonged to the lordship marcher of
Monmouth, was an isolated fragment of Monmouthshire
down to 1845; on the other hand a strip on the east
side of the Grwyne Fawr called the Ffwddog belonged
to Herefordshire till 1893[1], and the same county also
claimed a spot called in old maps Crooked Billet, lying
between Trellech Grange and the Devauden.

[1] It still goes with Herefordshire for parliamentary purposes.

4. Surface and General Features.

The surface of the county may be conveniently considered in five divisions :—the eastern hills, the "Levels," the central plateau, the coal district of the west, and the mountains of the northern portion.   We will take them in order.

The eastern hills, which are really a continuation of the Forest of Dean system, are a wild, wooded, and heathy tract ranging from 400 to 1000 feet in height, and comprising Penallt, Trellech, Llanishen, Chepstow Park, Earlswood and Wentwood. From Penallt to Windcliff, a rocky height of 770 feet, these hills rise abruptly from the Wye, but here they fall away from the river in a westerly direction till they terminate at Christchurch, overhanging Caerleon. At Trellech Beacon, now thickly planted with firs, they reach 1003 feet, at Chepstow Park near Devauden 932, and at the top of Wentwood 989 feet. Conspicuous spurs to the south of Wentwood are Mynydd Llwyd (Grey hill) 901, and Mynydd Allt Tir Fach 791 feet.   Here and there minor streams find their way southward to the Channel, but the north-western face of the lower extremity of this range descends steeply to the Usk, reminding us on a small scale of the western escarpments of the Chilterns and the Cotswolds.

The "Levels" lie to the south between the South Wales railway and the Severn Sea. The only flat ground of any extent in the county, they are divided into two parts by the mouth of the Usk, the part between that river and Portskewet being known as Caldicot Level, and that between the Usk and the Rhymney as Wentllwg, properly Gwynllwg (Wenllwch), Level. Locally they have long been known as "the Moors," and they contain some of the deepest and richest soil in the county. They are drained by wide ditches called "rheens" or "reens," which are crossed with the aid of a long jumping pole or "powt." As we shall explain in a later chapter, these moors are protected from the incursions of the sea by a stone wall.

The central plateau occupies the interior of the county between the Usk and the eastern hills. It is an undulating tract with many lesser elevations from 250 to 650 feet in height, intersected by numerous brooks and covered with copses, cultivated fields, orchards, and meadows. The soil is remarkably fertile and all kinds of crops can be easily grown. In the centre is Raglan, with its once splendid mansion ; its chief town is Usk, charmingly situated on the river of that name, but marred by an ill-placed railway bridge. Towards the north this plateau gradually rises till it culminates in the Scyrrid Fawr (1601 feet), Campstone Hill (884 feet), and the Graig (1389 feet). On the further side these heights overlook the valleys of the Honddu and the Monnow.

The western mountains containing the coal measures would be bisected by a diagonal line drawn from Abergavenny to the Rhymney at Bedwas. They form the eastern extremity of the great mountain chain which stretches from the Towey to the Usk, and in our county are penetrated by four valleys—not counting that of the

Rhymney—the waters of which all ultimately find their way into the Usk. They range from 900 to 1900 feet in height—the Blorenge being 1834 feet, and the Coity mountain to 1905 feet—and are the home of the great mining industries of the county; it is not surprising therefore that their population including that of Newport largely exceeds that of all the rest of the county put

The Blorenge from the East

together. This state of affairs, however, dates only from the end of the eighteenth and beginning of the nineteenth centuries. Before that time these hills and valleys were as sparsely populated as the district to the north of the Usk.

Our last division, the northern mountain portion, comprises as much of the Black Mountains as falls within the county boundary, that is the Mynydd Pen-y-fal (Sugar-loaf) 1955 feet, and the projection already mentioned containing Cwmyoy and Llanthony. In the extreme north corner of this the backbone of the ridge reaches 2091 feet, the highest point in the county.

5. Watershed and Rivers.

The principal rivers of the county are the Wye and the Usk, both of which flow into the Severn Sea, their mouths being fifteen miles apart. Their respective basins are divided by a watershed so narrow that the sources of their tributaries are sometimes within a few yards of each other. Let us trace the line of this watershed on a good map, the scale of which should not be less than one mile to the inch.

If we start in the north-west with the ridge of Chwarel-y-Fan and Bal-mawr, we find that it separates the Grwyne Fawr on our right from the Honddu on our left, the former stream belonging to the basin of the Usk, and the latter to that of the Wye. Proceeding in a south-easterly direction, the line of the watershed takes us along the flank of Bryn Arrw and across the railway close to Llanfihangel Crucorney station where the source of the Gavenny—an Usk tributary—is close to that of a streamlet flowing north into the Honddu, and then round the eastern side of the Scyrrid Fawr to Llanddewi Rhydderch : thence between the Pant and Ffrwd brooks to Penrhôs, Tregare, and Pen-y-clawdd, and so up Cwmcarfan hill, the spur of which divides the brooks flowing into the Trothy and the

Olwy respectively. We now pass round Trellech, and keep along the western edge of the hills to Llanishen, near which place we pass the head of the Angidy, the last stream which flows to the Wye, and so reach the Devauden, where the line of the watershed trends westward, and passing Newchurch follows the top of Wentwood till it terminates at Christchurch : in this part of

Fiddlers Elbow, River Wye, near Monmouth

its course the streams on its northern side still flow down to the Usk, while those on its southern side make directly for the Severn Sea.

We may now consider the two great rivers themselves. The Wye rises on Plynlimon, and after a devious course through the counties of Montgomery, Brecon, Radnor, and Hereford, first touches Monmouthshire two miles above

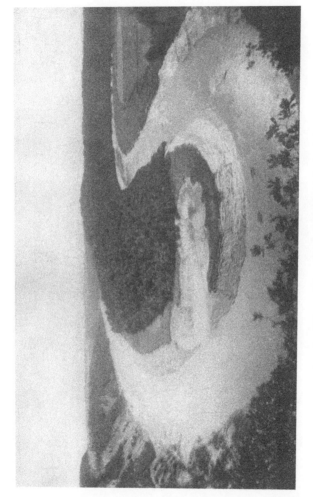

On the Wye: Piercefield Woods at low water

E. M.

2

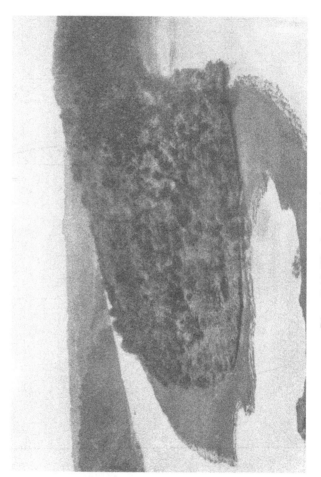

Piercefield Woods at high water

Wyastone Leys, where we began our perambulation of the
boundaries (Chapter 3). Hence it forms the boundary of
the county for a couple of miles, dividing it from Herefordshire, and then flows through it, passing the town of
Monmouth on its right bank, as far as Redbrook, when it
again becomes the boundary for the rest of its course,
dividing us now from Gloucestershire. The weirs or
rapids are numerous, for instance Bigsweir, Brockweir (the
weir at the brook), Lynweir (the weir by the maple tree),
Ashweir and Plumweir (the weirs by the trees of these
names), Stoneweir (where the bed is not rock, but composed of small stones of the great moraine which here
crossed the valley), Wallweir (the weir near the spring,
wylle), Baddyngesweir, or Hookweir (where the salmon
were caught by a *bait* : this has given its name to Banager
cliffs on the Gloucestershire side), and Troughweir (where
the salmon were caught in a trough or trap). The steep
cliffs that flank the stream on either side are clothed with
wood ; in some places they rise from the water's edge, in
others they leave it fringed by a strip of green pasture.
Between Tintern and Chepstow the course of the river is
remarkably sinuous, the hard limestone rocks which it
first encounters opposite the Abbey, and which crop up
now on one bank and then on the other, forcing the stream
away from them in the easier direction. Thus after having
been swept westwards by the Banager cliffs on the left
bank, it receives from the rocks of Windcliff and then
from those of Piercefield on its right bank, first a southward and then an eastern and north-eastern impetus round
the softer substance of the Gloucestershire bank, which

On the Wye: the Lancaut Peninsula

thus becomes the remarkable peninsula of Lancaut. The limestone cliffs on either side then confine it in a straight line known as "the long reach," till the rocks at Chepstow again drive it eastwards towards those at Tutshill, which in their turn direct its course towards Hardwick, whose cliffs turn it eastward. Hewin's rock once more deflects it and so it enters the Severn, leaving Beachley on its left.

The Monmouthshire affluents of the Wye are (1) the Mallybrook, a streamlet which comes down from Welsh Newton across the Herefordshire border, and comes in at Dixton : (2) the Monnow, which rises in the Black Mountains near Craswall, turns to the north-east at Allt-yr-ynys, and south-east at the Monmouth Cap : (3) the Trothy (Welsh Tro-ddi), which rising to the west of the Graig divides the hundreds of Skenfrith and Abergavenny, as after turning eastwards it divides the hundreds of Skenfrith and Raglan, and passing Llantilio Crossenny, Dingestow, and Michel Troy, joins the Wye half a mile below the Monnow : (4) Whitebrook, which comes in a mile and a half above Bigsweir : (5) the brook which produces the Cleiddon Falls in the steep gorge above Llandogo, and (6) the Angidy, which flows through a steep wooded gorge and joins the Wye at Tintern. The first three are well above the tidal portion of the river, which for practical purposes may be described as extending up to Llandogo, but the waters of the Atlantic being contracted in their upward course by the narrows known as "The Shoots" at the New Passage, and being again headed back by the projecting rocks at Beachley,

Wyastone Leys

enter the Wye with great velocity, and at spring tides
their influence is felt as high as Bigsweir.

The difference between high and low water mark is
always considerable : at a spring tide it has been known
to reach 49 feet above the lowest known ebb level, but
the average would be about 45 feet.   In the March
and September springs, a bore is formed in the Long
Reach resembling the famous Severn bore, but on a
smaller scale.   The flood after passing Chepstow bridge
is ponded up under the castle rock by a spit of firm
ground, which may be seen at low water projecting from
the Gloucestershire bank.   At length this spit is covered,
and the body of water rushes on to be contracted where
the channel narrows just above the old Roman crossing
which will be noticed in a later chapter.   It then mounts
up and advances up the reach with a head about two feet
high, and a roar that may be heard on the cliffs above.

The tour down the Wye from Ross to Chepstow—
which was necessarily taken by water, there being no
good roads in the valley at that time—was first brought
into fashion in the middle of the eighteenth century by
John Egerton, afterwards Bishop of Durham, who was
appointed to the rectory of Ross in 1745, and continued
to reside there occasionally up to the time of his death in
1787.   The poet Gray was one of those who made the
tour, and its beauties were described in a patronising style
by that professed admirer of the picturesque, William
Gilpin, and also by Francis Wheatley the artist, but it
is not of these that visitors who make the tour to-day will
think ; they will only remember that these scenes drew

from Wordsworth the "Lines composed a few miles above Tintern Abbey," when he revisited the Wye with his sister in the summer of 1798.

Passing from the Wye to the Usk we may note that the estuaries of the little brooks which come down from the hills about Newchurch and Wentwood to the Severn Sea between the two rivers are called pills (Welsh *pwll*) ; thus we have St Pierre pill at the mouth of the Meurig, Caldicot pill at the mouth of the Nedden or Troggy, and Magor pill at the mouth of the little stream that rises above Penhow.

The Usk rises at the foot of the Carmarthenshire Van, and after a course of some forty miles through Breconshire, reaches our county about halfway between Crickhowell and Abergavenny at the point where it receives the Gwenffrwd on its left bank. At Abergavenny it turns southward, past Llanelen, and skirting the foot of the mountain district on its right takes its sinuous way through pleasant pastures to the town of Usk. Thence having the heights of Wentwood on its left it reaches Caerleon and then Newport, three miles below which town it joins the Severn Sea. The whole of its course through Monmouthshire is very beautiful, and there are numerous villages and country houses lying on its banks, features for which the more confined valley of the Wye has less scope.

As for the tributaries of the Usk, those on the left bank are mere brooks, but those on the right bank embrace all the streams that drain the Monmouthshire coalfield excepting the frontier river, the Rhymney. To begin with the former, the Gwenffrwd, like the Nant Iago a little below it, is a mountain torrent which rises on the summit of the Sugar-loaf, and in its short course of three miles descends some 1700 feet; the Gavenny and its affluents drain the valley which descends from Llanfihangel Crucorney to Abergavenny; the Ffrwd comes in near Clytha; and the Olwy descending from Trellech, and

Llanelen Bridge

joined by an affluent from Devauden called the Pill Brook, comes in below the town of Usk.   On the right bank the Berthin and the Sor are small brooks which flow in the one just above Usk, and the other just above Caerleon. Half a mile below the confluence of the Sor the Usk is joined by a more considerable stream, the Afon Lwyd, which rising in the hills at Blaenavon passes Abersychan,

Pontypool and Lantarnam, and skirts Caerleon on the north-east.

Leaving the canal for a later chapter, two miles below Newport we come to the mouth of the Ebwy (Ebbw) which winds like a snake through the alluvial district below the hills. We may follow its course upwards

On the Ebbw River

through Tredegar Park to Risca, where it is joined by the Sorwy (Sirhowy), and by Abercarn under the Crumlin viaduct to Aberbig, where it divides into two branches ; that to the east, called the Ebwy Fach, ascends to Abertillery—the confluence as its name indicates of the Tillery (Tilerau)—and so to Blaina; that to the west, the

Ebwy Fawr, to the modern mining settlements of Ebbw Vale and Beaufort. If we return to the confluence of the Sorwy, we shall find that it conducts us up a narrow valley which separates the ancient parishes of Mynyddis-lwyn and Bedwellty, the churches of which lie back on the hilltops, to Tredegar and Dukestown, both centres of a mining population.

Lastly we have the Rhymney, which divides us from Glamorgan.   On its left bank is the mining town to which it gives its name, and a foreign-sounding settle-ment called from the sign of a public house Fleur-de-lis. Near Bedwas, a mile or so north of Caerphilly, it makes a bend to the east, but at Upper Machen it resumes its southward course till it reaches the Bristol Channel.   Its whole course is from 30 to 40 miles in length.

6. Geology and Soil.

By Geology we mean the study of the rocks, and we must at the outset explain that the term *rock* is used by the geologist without any reference to the hardness or compactness of the material to which the name is applied; thus he speaks of loose sand as a rock equally with a hard substance like granite.

Rocks are of two kinds, (1) those laid down mostly under water, (2) those due to the action of fire.

The first kind may be compared to sheets of paper one over the other. These sheets are called *beds*, and such beds are usually formed of sand (often containing pebbles), mud or clay, and limestone, or mixtures of these materials.

They are laid down as flat or nearly flat sheets, but may afterwards be tilted as the result of movement of the earth's crust, just as we may tilt sheets of paper, folding them into arches and troughs, by pressing them at either end. Again, we may find the tops of the folds so produced worn away as the result of the wearing action of rivers, glaciers, and sea-waves upon them, as we might cut off the tops of the folds of the paper with a pair of shears. This has happened with the ancient beds forming parts of the earth's crust, and we therefore often find them tilted, with the upper parts removed.

The other kinds of rocks are known as igneous rocks, and have been melted under the action of heat and become solid on cooling. When in the molten state they have been poured out at the surface as the lava of volcanoes, or have been forced into other rocks and cooled in the cracks and other places of weakness. Much material is also thrown out of volcanoes as volcanic ash and dust, and is piled up on the sides of the volcano. Such ashy material may be arranged in beds, so that it partakes to some extent of the qualities of the two great rock groups.

The production of beds is of great importance to geologists, for by means of these beds we can classify the rocks according to age. If we take two sheets of paper, and lay one on the top of the other on a table, the upper one has been laid down after the other. Similarly with two beds, the upper is also the newer, and the newer will remain on the top after earth-movements, save in very exceptional cases which need not be regarded by us here, and for general purposes we may regard any bed or set of beds resting on any other in our own country as being the newer bed or set.

The movements which affect beds may occur at different times. One set of beds may be laid down flat, then thrown into folds by movement, the tops of the beds worn off, and another set of beds laid down upon the worn surface of the older beds, the edges of which will abut against the oldest of the new set of flatly deposited beds, which latter may in turn undergo disturbance and renewal of their upper portions.

Again, after the formation of the beds many changes may occur in them. They may become hardened, pebble-beds being changed into conglomerates, sands into sand-stones, muds and clays into mudstones and shales, soft deposits of lime into limestone, and loose volcanic ashes into exceedingly hard rocks. They may also become cracked, and the cracks are often very regular, running in two directions at right angles one to the other. Such cracks are known as *joints*, and the joints are very important in affecting the physical geography of a district. Then, as the result of great pressure applied sideways, the rocks may be so changed that they can be split into thin slabs, which usually, though not necessarily, split along planes standing at high angles to the horizontal. Rocks affected in this way are known as *slates*.

If we could flatten out all the beds of England, and arrange them one over the other and bore a shaft through them, we should see them on the sides of the shaft, the newest appearing at the top and the oldest at the bottom,

much as in the table annexed. Such a shaft would have a
depth of between 10,000 and 20,000 feet. The strata
beds are divided into three great groups called Primary
or Palaeozoic, Secondary or Mesozoic, and Tertiary or
Cainozoic, and the lowest Primary rocks are the oldest
rocks of Britain, and form as it were the foundation stones
on which the other rocks rest. These are usually termed
the Pre-cambrian rocks. The three great groups are
divided into minor divisions known as systems. The
names of these systems are arranged in order in the table.
On the right hand side, the general characters of the rocks
of each system are stated.

With these preliminary remarks we may now proceed
to a brief account of the geology of the county.

This, perhaps, will be best understood by making a
beginning in the central part. Here an axis of elevation
(or anticlinal), approximately from south-west to north-
east, assisted by the denudation of the valleys, has exposed
the Upper Silurian Rocks on the surface in two areas.
The first, of small dimensions, is at the extreme south-
west of the county, along the estuary of the Rhymney,
where the Ludlow Shale and the Wenlock Limestone
and Shale can be observed. The other, an elliptical area
about nine miles by four, extends from near Llanfrechfa
to a point east of Llanfair Kilgiden. In this area the
same rocks are successively exposed; the lowest in the
series being seen at Cilfigan.

This axis of elevation has left its most northern opera-
tion in the hills about Grosmont; though, from insufficient
denudation, the Silurian is not there exposed.

Names of Systems	Subdivisions	Characters of Rocks
TERTIARY		
Recent / Pleistocene	Metal Age Deposits Neolithic ,, Palaeolithic ,, Glacial ,,	Superficial Deposits
Pliocene	Cromer Series Weybourne Crag Chillesford and Norwich Crags Red and Walton Crags Coralline Crag	Sands chiefly
Miocene	Absent from Britain	
Eocene	Fluviomarine Beds of Hampshire Bagshot Beds London Clay Oldhaven Beds, Woolwich and Reading Thanet Sands [Groups]	Clays and Sands chiefly
SECONDARY		
Cretaceous	Chalk Upper Greensand and Gault Lower Greensand Weald Clay Hastings Sands	Chalk at top Sandstones, Mud and Clays below
Jurassic	Purbeck Beds Portland Beds Kimmeridge Clay Corallian Beds Oxford Clay and Kellaways Rock Cornbrash Forest Marble Great Oolite with Stonesfield Slate Inferior Oolite Lias—Upper, Middle, and Lower	Shales, Sandstones and Oolitic Limestones
Triassic	Rhaetic Keuper Marls Keuper Sandstone Upper Bunter Sandstone Bunter Pebble Beds Lower Bunter Sandstone	Red Sandstones and Marls, Gypsum and Salt
PRIMARY		
Permian	Magnesian Limestone and Sandstone Marl Slate Lower Permian Sandstone	Red Sandstones and Magnesian Limestone
Carboniferous	Coal Measures Millstone Grit Mountain Limestone Basal Carboniferous Rocks	Sandstones, Shales and Coals at top Sandstones in middle Limestone and Shales below
Devonian	Upper } Mid } Devonian and Old Red Sand- Lower } stone	Red Sandstones, Shales, Slates and Lime- stones
Silurian	Ludlow Beds Wenlock Beds Llandovery Beds	Sandstones, Shales and Thin Limestones
Ordovician	Caradoc Beds Llandeilo Beds Arenig Beds	Shales, Slates, Sandstones and Thin Limestones
Cambrian	Tremadoc Slates Lingula Flags Menevian Beds Harlech Grits and Llanberis Slates	Slates and Sandstones
Pre-Cambrian	No definite classification yet made	Sandstones, Slates and Volcanic Rocks

South-east and north-west of this axis the rocks newer than the Silurian dip in either direction.

The dip to the eastward is the more gentle, with the result that the whole surface of the county on that side of the axis, with the exceptions to be mentioned, is of the Old Red Sandstone formation. This however in the south-east part of the county passes under a narrow strip of Carboniferous Limestone of which the margin or outcrop is first seen at Magor, where this rock reappears after passing under the Severn Sea from Portishead Point in Somerset. The outcrop then passes north in the direction of the Grey Hill, skirts the southern flanks of that and Shirenewton Hills, and then trends in the direction of Chepstow Park, from the foot of which, by a fault, it is thrown eastward to a point a little east of St Arvans. Thence passing at the west or back of the Windcliff, and taking in Penrose Hill, it turns sharply south-east and leaves the county about a mile below Tintern Abbey. Between this line of outcrop and Severn and the Wye this part of the county (subject to the overlies to be presently mentioned) is wholly overlaid by the limestone; this strip of which, having crossed the Wye valley, becomes still narrower under the ridge of Tidenham Chase in Gloucestershire, ultimately expanding in that county to pass under and take in the Dean Forest coalfield.

An outlier of lower Carboniferous shale on Penallt common near Trellech is the only relict of the enormous denudation of Carboniferous rocks to which the Old Red Sandstone area we have considered was subjected.

The Severn Sea lies in a great "trough fault," nearly parallel to the axis of elevation above mentioned. This on its south-east side, in Gloucestershire, along the Ridge Hill near Patchway, has brought down, or left, the Trias resting against the Carboniferous Limestone. On the Monmouthshire side it has left the Pennant Sandstone of the Bristol coalfield faulted against the Carboniferous Limestone. This Pennant Sandstone is exposed in the Charstone Rock, Lady Bench, and Gruggy at low water.

Over the limestone area already considered sporadic patches of dolomitic conglomerate are to be found in the neighbourhood of Chepstow and Magor. The newer beds of the Trias have partly refilled the valleys formed by the small streams running to the Severn, such as the Pool-meyric and Nedden brooks, and they flank and underlie the low grounds of Caldicot Level. Superimposed on this near Llanwern is a cap of Lias (resting on a thin bed of Rhaetic), which also appears on the shore at Goldcliff. A very small outlier of these beds also remains on the bank of the Ebbw near Tredegar Park.

Returning now to the central axis of elevation, we find the beds dipping much more sharply towards the north-west. In this direction, within a few miles at the most, we lose the Old Red Sandstone as it passes under the Carboniferous Limestone along the flanks of the Mynydd Machen, the Mynydd Maen, and the Blorenge succes-sively. At the latter point the outcrop of the limestone which is exposed in a very narrow belt turns westward along the Usk Valley, and soon leaves the county. The Sugar-loaf and the parts of the Black Mountains

Abergavenny: the Sugar-loaf from the East

within the county are all of the Old Red Sandstone formation.

The limestone immediately passes westward under the Coal Measures of the South Wales system, which are, for the greater part, exposed along the upper parts of the range of hills last mentioned ; and continue to, and far beyond, the county boundary in the Rhymney Valley.

In this area 23 seams of coal are recognised corresponding to the same number of seams in the Dean Forest coalfield east of the Wye. This fact, even if other evidence were wanting, would alone indicate the vast amount of denudation that the central part of the county has been exposed to in geological times. To appreciate this we must in imagination restore over the Silurian at Usk first the thickness of the Old Red Sandstone, at least 2500 feet, and then carry over this, as in an arch, from Pontypool to the Dean Forest, the limestone of about 800 feet, and the Coal Measures of 2500 feet more ; and we then have to estimate, if we can, the forces and time required to plane down this mighty mass, and finally plough it out into the valleys that adorn the county.

7. Natural History.

Various facts, which can only be shortly mentioned here, go to show that the British Isles have not existed as such, and separated from the continent, for any great length of geological time. Around our coasts, for instance, are in several places remains of forests now sunk beneath the sea, and only to be seen at extreme low water.

3—2

Between England and the continent the sea is very shallow, but a little west of Ireland we soon come to very deep soundings. Great Britain and Ireland were thus once part of the continent, and are examples of what geologists call recent continental islands. But we also have no less certain proof that at some anterior period they were almost entirely submerged. The fauna and flora thus being destroyed, the land would have to be restocked with animals and plants from the continent when the land again rose, the influx of course coming from the east and south. As, however, it was not long before separation occurred, not all the continental species could establish themselves. We should thus expect to find that the parts in the neighbourhood of the continent were richer in species and those furthest off poorer, and this proves to be the case both in plants and animals. While Britain has fewer species than France or Belgium, Ireland has still less than Britain.

The fauna and flora of Monmouthshire are very extensive—more so than those of many a larger county. This is due to the diversity of surface and geological formation previously described. Mountain, marsh, woodland, rock, bog, and sea-coast—all furnish their characteristic species. In the rivers the otter is found—and hunted. The rocks and woods are the home of the badger and the fox. The former animal was in old days baited, and the latter, in spite of the natural difficulties of the country, is still hunted. The hounds in use, at any rate in the south of the county, are small and wire-haired—something between the large hound of the Midlands and the otter-hound. They are well suited for work in the steep hillsides, the tangled copses, and the rocky ground with which the county abounds. Four packs divide it between them—the Monmouthshire in the north, Lord Tredegar's in the south-west, and the Llangibby and Itton packs in the south. In so intricate a country, unless he can be driven out into the open, the fox is as often lost as killed. The rocky woods along the Wye in particular abound in foxes, which are rarely dislodged.

The larger birds of prey such as the kite, which less than a century ago still bred in the county, have mostly disappeared, but the buzzard is occasionally seen and the raven still exists, as do the peregrine, hobby, and merlin. Many are kept down by the gamekeeper. The consequences are the same as elsewhere, and the farmer's enemies, such as woodpigeons, sparrows, starlings, rats, and mice abound. No list of the birds of the county has been published, but if compiled it would be a long one. Along the coast would be found the commoner shore birds, such as sandpipers, plovers, stints and dunlins, gulls and ducks, especially the handsome sheld-duck. The heron, once abundant on the Wye, is destroyed by the lessees of the fisheries, but still frequents the levels and the valley of the Usk; the bittern is not unknown, the snipe is common on the moors, and the woodcock more than usually abundant in the boggy woods. All the common inland birds are found, and among the rarer ones the pied fly-catcher, cirl bunting, and spotted crake. Some of the smaller species have, however, decreased in numbers during the last dozen years or so, largely owing, it is thought, to

the plundering of their nests by the magpie, a bird which
has enormously increased in some parts during this period.
The ring-ousel and dipper breed, and the grouse occurs
in the mountains of the north-west.

Monmouthshire is a famous county for butterflies : in
the open spaces of the woods the Vanessas and the Fritil-
laries are fond of sunning themselves, while here and there
that characteristic butterfly of the southern border counties,
the Comma, may be seen flitting about the undergrowth.

Monmouthshire has produced no great naturalist, or
it would be singular that a county so rich in its fauna and
flora has never been treated as a whole.   No book on the
beasts, birds, or insects of the county has ever been written,
and though more attention has been paid to the plants,
we are still without a complete flora.   The number of
species is very extensive.   Very few Natural Orders, and
but a small proportion of genera, are unrepresented, and
the eastern side of the county alone contains more than
820 species of flowering plants, and 21 species of fern.
This is an exceptional amount for so small an area.
Among the rarer species may be mentioned *Aconitum
napellus, Cardamine impatiens, Pyrus rotundifolia, Sedum
rupestre, Crithmum maritimum, Rubia peregrina, Cam-
panula patula, Orobanche hederae, Calamintha parviflora,
Polygonum mite, Euphorbia stricta, Cephalanthera ensifolia,
Carex digitata*, and *Carex montana*.   The rocky woods on
the Wye are a magnificent sight in spring and again in
autumn owing to the many kinds of trees which they
contain, all contrasting with the dark evergreen of the
yew, which is here both indigenous and abundant.

8. The Severn Sea.

In prehistoric times the Bristol Channel, or Severn Sea, which forms the southern boundary of our county, was a marsh through which the river Severn found its way to the ocean ; but when at a later period, still prehistoric, the land subsided, and the sea flooding the marsh pushed its way up as far as Gloucester and even beyond, the appearance of the estuary gradually became something like what it is at the present day. We must not be more precise, because there is reason to believe that even in historic times the shoals have shifted and altered the course of the river. The land itself too—the grassy mud-flats covered at high tides—formerly reached farther out than at present, for you may see signs everywhere of the encroachment of the sea, and particularly at the two extremities of Caldicot Level—Sudbrook on the east, where half of the ancient camp that defended the landing place has been washed away, and Goldcliff on the west, where the front of the cliff has been cased with strong masonry to prevent further inroads. Both these places have an historical interest: at Sudbrook was an alternative landing-place to Porthiscoed, now St Pierre pill, from which the later village of Portskewet takes its name, and Goldcliff, hard by which was a Benedictine Priory, is an isolated limestone rock imposed upon a sandstone base. The latter is full of yellow mica, so that the whole cliff seaward, when the sun was upon it, shone like gold. This effect however can no longer be seen owing to the casing of masonry just mentioned.

A glance at a good map will show that the most dangerous part of the Channel is that opposite Sudbrook, beneath which the Severn Tunnel now runs. On the English side are the extensive reefs called the English Stones, and on the Welsh side lie Charstone rock, Lady Bench, Gruggy (Welsh *Crugiau*, mounds), Mixoms, and the Black Bedwins—all uncovered at low water. Between these reefs, English and Welsh, is a narrow strait, called the Shoots, the only passage at low tide. Further west on the Welsh side are the shoals known as Bedwin Sands, Porton Grounds, Welsh Grounds, and Usk Patch, while in mid-channel is a rock, uncovered at the highest tides and conspicuous far and wide, called the Denny. There was on the Monmouthshire coast, perhaps even in historic times, a forest of hazel and oak, now all submerged, of which remains may be seen in the mud-banks at Goldcliff.

In ancient times the encroachments of the sea at high tide came in far beyond the mud-flats, and inundated that wide strip of country reaching to the base of the hills which is known as the Levels, or the Moors. Before they were reclaimed from the sea these must have resembled a vast lagoon, impracticable for all purposes of pasture or cultivation. It was the Romans who first saw their potential value, and they therefore constructed at a short distance inland from the sea a strong rampart twenty miles in length stretching from Sudbrook to the Usk and from the Usk to the Rhymney. The moors were then drained by wide ditches, now called "reens," having outlets in the embankment guarded by sluices or "gouts."

Now even without direct evidence we should be disposed to attribute this long rampart to the Romans, for an effort of engineering on so vast a scale was far beyond the scope of early inhabitants of the country, and it was only when the Pax Romana had paved the way that such operations became possible. But this is not the only kind of reasoning upon which we are able to rely. In the autumn of 1878 an inscribed stone was washed out of the soil at a depth of five or six feet on the seaward side of the bank at Goldcliff. The letters cut on this stone clearly relate to the century of Statorius forming part of the First Cohort. If the third line be interpreted M[uri] M[ille] i, the whole would mean that the century of the First Cohort commanded by Statorius built or repaired one mile of the embankment.

The embankment thus constructed was of earth with wattling on the seaside, which gradually gave way to a stone wall, and though able to keep out the water at ordinary times, the sea occasionally washed over it at high spring tides, causing considerable damage, for the land within lies from 15 to 20 feet below the level of the tide, and under these circumstances is easily flooded. The worst flood on record occurred in January 160$\frac{8}{9}$ when the spring tide, backed by a strong south-west wind, swept over the embankment and buried the whole country. In the churches of St Bride's, Gwynllwg (Wenllwch), and of Goldcliff are contemporary brass plates commemorating the disaster, and recording the depth of the flood. There is reason to believe that the population of the Moors was at that time much more numerous than it now is, and the destruction of life and property, according to a pamphlet printed at the time, was very great: "The number of men drowned are not known to exceed 2000"—a very safe statement; further "all wild beasts and vermin tried to escape from the water by getting to the most elevated banks and parts of the land, where were collected dogs, cats, moles, foxes, hares, conies, rats, mice in abundance, and, strange to say,

Usk Lighthouse, near Newport

the one of them never offered to annoy the other, but in gentle sort freely enjoyed the liberty of life." Another inundation, though on a smaller scale, occurred towards the end of the eighteenth century. Moreover the bank at high tide was constantly exposed to erosions, and it therefore became necessary to case it on the outside with masonry, so that it is now a veritable stone wall. The whole of this sea wall, together with the drainage of the Moors, is managed and kept in order by the Court of Sewers.

From what has been said it will be readily understood that the navigation of the Severn Sea is dangerous and difficult. Pilot boats sometimes go down as far as Lundy on the look out for vessels. Our Monmouthshire lighthouses are three, one on Charstone rock and one on either side of the mouth of the Usk. Lastly, there is one curious point which must not pass unnoticed, and that is that the parish boundaries run out from the shore in straight lines to the county boundary, which is the deep water channel, or the centre line at low water, thus serving to define the rights of fishing and of wreck appurtenant to the ancient lordships.

9. Climate.

The climate of a country or district is, briefly, the average weather of that country or district, and it depends upon various factors, all mutually interacting, upon the latitude, the temperature, the direction and strength of the winds, the rainfall, the character of the soil, and the proximity of the district to the sea.

The differences in the climates of the world depend mainly upon latitude, but a scarcely less important factor is proximity to the sea. Along any great climatic zone there will be found variations in proportion to this proximity, the extremes being "continental" climates in the centres of continents far from the oceans, and "insular" climates in small tracts surrounded by sea. Continental climates show great differences in seasonal temperatures, the winters tending to be unusually cold and the summers unusually warm, while the climate of insular tracts is characterised by equableness and also by greater dampness. Great Britain possesses, by reason of its position, a temperate insular climate, but its average annual temperature is much higher than could be expected from its latitude. The prevalent south-westerly winds cause a movement of the surface-waters of the Atlantic towards our shores, and this warm-water current, which we know as the Gulf Stream, is the chief cause of the mildness of our winters.

Most of our weather comes to us from the Atlantic. It would be impossible here within the limits of a short chapter to discuss fully the causes which affect or control weather changes. It must suffice to say that the conditions are in the main either cyclonic or anticyclonic, which terms may be best explained, perhaps, by comparing the air currents to a stream of water. In a stream a chain of eddies may often be seen fringing the more steadily-moving central water. Regarding the general north-easterly moving air from the Atlantic as such a stream, a chain of eddies may be developed in a belt parallel with its general direction. This belt of eddies or cyclones, as they are termed, tends to shift its position, sometimes passing over our islands, sometimes to the north or south of them, and it is to this shifting that most of our weather changes are due. Cyclonic conditions are associated with a greater or less amount of atmospheric disturbance ; anticyclonic with calms.

The prevalent Atlantic winds largely affect our island in another way, namely in its rainfall. The air, heavily laden with moisture from its passage over the ocean, meets with elevated land-tracts directly it reaches our shores—the moorland of Devon and Cornwall, the Welsh mountains, or the fells of Cumberland and Westmorland —and, blowing up the rising land-surface, parts with this moisture as rain. To how great an extent this occurs is best seen by reference to the accompanying map of the annual rainfall of England, where it will at once be noticed that the heaviest fall is in the west, and that it decreases with remarkable regularity until the least fall is reached on our eastern shores. Thus in 1906, the maximum rainfall for the year occurred at Glaslyn in the Snowdon district, where 205 inches of rain fell; and the lowest was at Boyton in Suffolk, with a record of just under 20 inches. These western highlands, therefore, may not inaptly be compared to an umbrella, sheltering the country further eastward from the rain.

The above causes, then, are those mainly concerned in influencing the weather, but there are other and more local factors which often affect greatly the climate of a place, such, for example, as configuration, position, and soil. The shelter of a range of hills, a southern aspect, a sandy soil, will thus produce conditions which may differ greatly from those of a place—perhaps at no great distance—situated on a wind-swept northern slope with a cold clay soil.

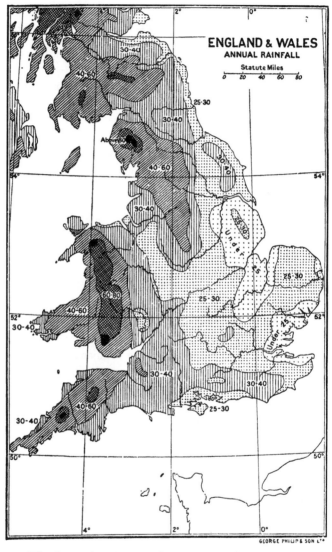

ENGLAND & WALES
ANNUAL RAINFALL
Statute Miles
0 20 40 60 80

30-40

40-60

25-30

30-40

30-40

Aber..60

40-60

30-40

30-40

Under 25

25-30

60-80

40-60

25-30

25-30

30-40

Under 25

30-40

30-40

30-40

40-60

30-40

25-30

GEORGE PHILIP & SON L.º

(The figures give the approximate annual rainfall in inches)

The character of the climate of a country or district influences, as everyone knows, both the cultivation of the soil and the products which it yields, and thus indirectly as well as directly exercises a profound effect upon Man. The banana-nourished dweller in a tropical island is of different fibre morally and physically from the inhabitant of northern climes who wins a scanty subsistence from the land at the expense of unremitting toil. These are extremes; but even within the limits of a county, perhaps, similar if smaller differences may be noted, and the man of the plain or the valley is often distinct in type from his fellow of the hills.

Very minute records of the climate of our island are kept at numerous stations throughout the country, relating to the temperature, rainfall, force and direction of the wind, hours of sunshine, cloud conditions, and so forth, and are duly collected, tabulated, and averaged by the Meteorological Society. From these we are able to compare and contrast the climatic differences in various parts.

In Monmouthshire the chief distinction is between the mountains of the west and north-west and the rest of the county. In the rainfall map it will be seen that the mountains lie in the 40—60 inches average annual rainfall division, while the country to the east of them lies in the 30—40 inches district. At Abergavenny, which lies at the foot of the mountains just where the Usk coming down from Breconshire bends to the south, the average rainfall for thirty years from 1870 was 37·82 inches, and for the 47 years 1864–1910 was 37·10

inches. In this period the highest rainfall in one year
was 52·2 inches in 1872; 1910 came next with 50·28
inches. The lowest was in 1864 when it was only
27·18 inches. In 1908, a dry year, the total rainfall at
Abergavenny was only 29·42, while higher up the Usk
valley, but outside the limits of our county, it was at
Crickhowell 35·33, at Glanusk 36·96, and at Talybont,
six miles below Brecon, 41·3. At Newport a ten years'
average for 1881–90 gave 40·38 inches. Most of these
figures are considerably higher than the average annual
rainfall for Great Britain, which is about 32 inches.

The mean annual temperature for most of the county
is between 48 and 50 Fahr., but in that part of the
county which lies along the Bristol Channel it is over
50 Fahr. It need hardly be said that, after a heavy fall,
the snow lies upon the hills long after it has disappeared
from the lower elevations. There are not sufficient
statistics available to justify any summary of the sunshine
recorded in the county.

10. People—Race, Language, Population.

The story of the peopling of Britain since it became
an island is a story of migrations. And these migrations
were ever westwards and north-westwards. Pressing on
from the barren steppes and dense forests of the east,
nation after nation passed onwards in search of new and
fertile settlements till the Atlantic, then the edge of the
world, forbade them to advance further. It was towards

the Atlantic accordingly that the first comers were always pushed, and when the migrations stopped, and the various batches of immigrants settled down in the eleventh century, the latest arrival would be found furthest to the east. From that time the history of our island has been one of amalgamation, sometimes forcible, sometimes voluntary, and of fixed settlement.

The first people to cross the Severn and the Dee were a short, dark-haired race akin to the Basques of north-western Spain,—the Iberian. They have bequeathed their peculiar features to a type common enough in the west of our island, and particularly in South Wales. The next comers were a tall, light-haired, blue-eyed race —the first swarm of the Celtic immigrations; they are now spoken of as Goidels, as distinguished from the Brythons who formed the second swarm of Celts. Driven onwards by these later comers the Goidels passed in large numbers over to Ireland and Scotland and became the ancestors of the people we term the Gaels. It need not, however, be supposed that the later comers wholly extirpated their predecessors ; the women, and the men of the old race who had been made the slaves of their conquerors, would remain, and gradually leaven the race. This was especially the case with the Iberians, who seem never to have been driven out, but to have remained in their settlements as a non-free population.

In Roman times the people of south-eastern Wales were known as Silurians (Silures)—a word of unknown origin. A proper name from the same root, Silulanus (or Silvianus), occurs on a leaden tablet discovered in the

Roman camp at Lydney on the edge of the Forest of Dean,
where there was a temple of the Celtic sea-god Nodens.
The people of Gwynllwg (Wenllwch) and Gwent—that
is Monmouthshire—are therefore Silurians; Celts, in point

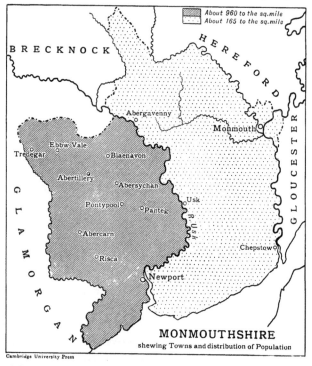

MONMOUTHSHIRE
shewing Towns and distribution of Population

Cambridge University Press

of fact, with a very large intermixture of Iberian. To-day
if we are English we call them Welsh, if Welsh Cymry.
There is of course in our county and in the rest of eastern
Wales a considerable sprinkling of English settlers. As

to language, east of the Usk it is now wholly English, though spoken by the natives with a marked Welsh intonation : west of the Usk in the hills it is still mainly Welsh, though English is understood. In point of fact, during the last hundred years Welsh has been steadily retreating westwards, but a century ago it might be heard in Chepstow market on any Saturday.

The population has considerably increased in the last half century. In 1851 it was only 157,418, or 273 to the square mile : in 1911 it was 395,778, or 724 to the square mile. This estimate would however be very deceptive if taken to represent the density of population in the county as a whole. Between the Wye and the Usk the houses are scattered and the population comparatively scanty, though the number of small farms and little freeholds which are sprinkled over hill and dale prevent it from being exclusively a county of villages of the English type. On the whole we shall not be far from the truth if we assign one-third of the whole population to the east of the Usk, and two-thirds to the west of that river.

11. Agriculture.

If we compare the face of the country in Wales with that of the purely Teutonic parts of England we are at once struck by this difference—in the former the farmhouses are scattered up and down the landscape apparently at haphazard, in the latter they are congregated in the villages. The reason for this is to be found far back

in the customs of the early settlers.   In Wales the tribe,
in England the village community formed the unit.
The tribe is the earlier form of union, consisting
as it does of a group of families related in blood, the
village community is a territorial unit held together by
the cultivation of the soil in common.   The Welsh at
the time of their settlement in Britain were still in the
tribal stage ; they were a pastoral rather than an agricultural people, and as each tribe effected its settlement
in any district its homesteads were spread over the whole
area, these scattered households combining together under
a steward in groups of twelve to fourteen to pay a food
rent of honey, cows, sheep, pigs, etc. to the tribal chief.
Afterwards, as population increased, fresh holdings were
carved out of the surrounding waste, and if these new
intakes encroached on the settlements of other tribes
tribal wars were the result.   Of this state of affairs
the present system of scattered dwellings is the lineal
descendant.   The village community was an institution
found from early times in the eastern parts of Britain,
and the village was a fixed settlement with definite limits,
round which were the arable or "common" fields, cultivated in strips of an acre or half an acre each, and
apportioned between the lord and his tenants.   After the
harvest these fields were thrown open as a common
pasture for the cattle of the cultivators.   The village had
also its open common for grazing ; and where possible its
meadows reserved for hay.   In Wales too, as in Scotland
and in Ireland, as regards arable husbandry the open field,
or to use an old expression the champion system of

cultivation, prevailed from early times, whether as an independent development, or as a common inheritance from the earliest tillers of the soil. And there is this peculiarity belonging to the system, whether found in connection with the village or with the tribal settlement, that the strips belonging to one holder are not contiguous, but scattered here and there in an apparently haphazard order over the field. It has been suggested that this may be due to the fact that as the common ploughing proceeded the strips were allotted in order to the several contributors to the common team, the first to the owner of the first ox, the second to the owner of the second ox, and so on: while the contributors of the ox-gear and the parts of the plough, coulter, share etc. would also get their strips. But whether this was the actual way in which the distribution was made or not, the end in view was clearly to equalise the value of the different holdings as far as possible. Had each man's strips lain together contiguously, one man might have had a very good holding, another a very poor one, owing to differences of quality in the soil of the same field. As it was each took his share of good and bad alike.

In England generally the parish church and the farmhouses are the village; in Wales the church is usually solitary, standing beside some stream, or up on the hillside, with a single farmhouse for its neighbour. The village, where it exists, is commonly at a distance from the church, and owes its origin to the need arising later for providing homes for the skilled artisan, such as the carpenter and the blacksmith, and for such agricultural labourers as had

families of their own and could no longer be accommodated in the farmhouse. Not that the number of mere
labourers is large ; the farms are small, and many of them
managed entirely by the farmer and his family. In
Monmouthshire the average number of labourers to each
farmer is now probably not more than one, while in
Gloucestershire it may be six or seven, and in Herefordshire four or five. The farms range from 30 to 150
acres ; 200 acres is regarded as quite a large holding,
and one of 300 is a rarity. In the last hundred and fifty
years the number of large landholders has increased,
but a considerable number of small freeholders remain,
particularly on the large enclosed commons, such as those
of Trellech and the Devauden. Many of them are
descended from squatters on the waste—a survival of the
old tribal outgrowth. Except in the case of such
enclosures as these, which took place about a century
since, the hedgerows were planted long ago and the fields
are small, as would be natural in a country of small freeholders, where the open field system would disappear
earlier than in England.

The conversion of arable into pasture land which
has been taking place everywhere during the last forty
years, is nowhere more marked than in Monmouthshire.
Out of a total of 347,600 acres there are now less than
40,000 under the plough—in fact roughly speaking one-
half of the county is grass, and of the other half one-
third arable, one-third woodland, and one-third mountain
and heath. From this it will be seen that woodmen,
shepherds, and cowmen constitute the greater part of

the labouring population. We may add that owing to the competition of the mining industries the rate of wages is comparatively a high one, ranging as it does from 17 to 19 shillings per week. The cultivation is generally the four field system—(1) roots, (2) barley and clover, (3) clover, (4) wheat. The best arable farms are in the south between Chepstow and Newport; in the north of the county a ploughed field stands out conspicuously among the mass of meadows and woods.

For the most part the farmer devotes himself to raising stock for the market, and for this purpose he finds nothing better than the long-horned Hereford breed; a few shorthorns he keeps to supply his family with milk, but it is only in the neighbourhood of the towns that regular dairy-farms exist. As for sheep, the smaller and hardier breeds are now in favour, as they do not require so much feeding on roots and hay in the winter time, and are excellent for mutton.

The extensive woods that clothe the steep banks of the Wye, as well as Wentwood, and the many coppices scattered over the face of the county, are generally cleared every fifteen years. The felling and cording of the undergrowth used to give constant employment to a number of men, and the hauling of it to the timber merchants, but now, owing to the difficulty of finding a market for the wood, the number of woodmen is considerably reduced, and in consequence, and also owing to the reduction in the number of farm labourers, hundreds of cottages in the eastern parts of the county are in ruins. Formerly a great quantity of timber and bark was

Sheep-washing on the Usk

carried down the Wye and exported from Chepstow, but this trade has long been extinct, and these articles are now trucked at the nearest railway station.

12. Industries and Manufactures.

The old domestic manufactures of Gwent, flannel and rough cloth, resembled those of the rest of Wales, and the enterprising town of Monmouth boasted a special feature in the woollen headgear known throughout the country as Monmouth caps, but the manufacture of these articles was long ago removed to Bewdley in Worcestershire, where as late as the beginning of the last century the caps were still sold under their old name. In Tudor times they were worn by soldiers, and it was in their Monmouth caps that the Welshmen at Crecy, according to Fluellen in Shakespeare's play of *Henry V*, wore the leek, which was then considered the emblem of Wales, though according to a recent writer this was an error for the daffodil. But the industries for which the county is most famous are not connected with wool but with iron. In this, as in other industries, the Romans were the pioneers, and both at Monmouth and at other places in the county abundant evidence of Roman smeltings has been found. From this time down to the days of the Tudors we hear no more of iron-works, but at that time the vast extent of wood in the county, as well as the facilities afforded by the Wye for the conveyance of heavy material, began to attract the attention of enterprising iron-masters. An abundant supply of wood was then and for the next two centuries an absolute necessity, for all smelting of the ore was done by charcoal, and not by pit coal.

It will be convenient to keep the iron-works on the Wye distinct from those in the west of the county, and to consider the former first. About 1570 a London company called "The Society for the Mineral and Battery Works" fixed upon Tintern as a suitable place for establishing a manufacture of iron wire. This was managed by a member of the Hanbury family, which afterwards settled at Pontypool. The distinguishing feature of their business was the introduction of wire-drawing by water power instead of man power, and for this they had a patent. At the same period an iron forge was set up in the neighbourhood as a venture independent of the wire-works, and both manufactures were continued with varying fortunes down to the last century. The iron-works, which were always worked with charcoal (it took 16 sacks of charcoal to make one ton of pig iron), came to an end in 1828, while the wire-works lasted for some fifty years longer.

At Pontypool iron ore began to be worked as early as the fifteenth century and in the next (in 1565) iron forges were established there by Capel Hanbury, but this Worcestershire family did not begin to reside there till the beginning of the eighteenth century. To these forges the ore found in other parts of the district was brought for smelting, but as the available wood became used up the works began to decline, nor was it till the discovery about 1760 that pit coal answered all the purposes of charcoal that they took a fresh start. It is from this period that the modern history of the South Wales iron industries may be said to begin.

In connection with the iron a japanned goods factory was set up at Pontypool in the reign of Charles II by a Northamptonshire family called Allgood, but it came to an end in 1822, though a branch survived at Usk till 1860. Other extinct industries are the paper mills formerly worked at Mounton in the Meurig valley near Chepstow, and also at Whitebrook; and the millstone factory at the breccia quarries near Trellech.

Among the industries not directly connected with iron may be mentioned the brattice-cloth works at Newport, a manufacture introduced into Wales about the middle of the last century. Brattice-cloth is a stout tarred cloth used for making partitions to secure the ventilation of coal mines, instead of wooden bratticing. The same firm afterwards introduced a plant for manufacturing rubber goods used in collieries, such as steam joints, valves, engine packings, hose, and belting.

But the staple industries of the county are the iron and tinplate manufactures. When once coal had taken the place of wood as the fuel employed for smelting, it was inevitable that all blast furnaces and foundries, which, like the Tintern works, were a long distance from the coal mines, should be closed, and now throughout the hill district there must be at least thirty different iron-works in blast. Outside this district and conveniently close to the Forest coalfield there is a tinplate factory at Redbrook, and there are also engineering works at

Cwmcarn, Abercarn

Chepstow, where such heavy manufactures as dock gates
and caissons are made.   The ore used in all these
furnaces is now imported from abroad, largely from
Spain.   Formerly it was dug out on the spot, but the
proportionate yield of metal is small, and little if any is
now worked.   Needless to say that these industries have
changed the whole character of the district : the valleys
are black with smoke, and the streams foul with impurities,
while the farms have disappeared from all but the upper
parts of the hills.   Here and there, however, the slopes
are still clothed with woods, above which the whitewashed
farmhouses are dotted over the higher ground.

Some idea of the present state of the iron trade at
Newport may be gained from the following figures for
the year 1908 :—

Rails   exported from Newport 19,632 tons.
Bar iron   „   „   1,210 „
Pig iron   „   „   97 „

There is one circumstance which must not remain
unnoticed in connection with the industrial development
of West Monmouthshire, and that is that, from the
earliest times to the present day, the pioneers, leaders,
and managers have seldom been native Welshmen, but
strangers from Scotland, from the north of England, or
from the manufacturing districts of the midlands.   If a
census were taken of the leading business men at
Newport to-day or of the great colliery company
managers in the hills it would be difficult to find a
genuine Welshman among them.

13. Mines and Minerals.

There is evidence that the Roman enterprise which developed the iron-industry in the Forest of Dean and in the east of Monmouthshire also extended to the west, and we have seen in the last chapter that a member of a Worcestershire family established iron-works at Pontypool in the time of Elizabeth; but it was not till the middle of the eighteenth century was past that the great changes began which have transformed these western valleys from a secluded sheep-walk into the home of a teeming industrial population. It was then that what we may call the English and Scottish invasion began, and that the mineral wealth which the native landowners showed no inclination to develop began to be exploited on a larger scale than ever before. Among the pioneers were Messrs Monkhouse, Fothergill & Co., Hill & Co., and Harford, Partridge & Co. The ore used by all these firms was the ironstone (called "Welsh mine") of the district and, as at Tintern, the fuel they first employed for smelting was charcoal. It was not long, however, before its place began to be taken by pit coal.

Coal-mining, then, in its inception was merely an auxiliary process to the iron-foundry. It was reserved for the steam engine to find another purpose for it, and to create a demand for Welsh coal which those who first used it little suspected. For the lower measures of the

South Wales coal[1] have never been a favourite for domestic use, consisting as they do almost entirely of carbon and only a very small proportion of bitumen; thus this coal is a kind of natural coke, which does not flame when ignited, and is therefore invaluable both for smelting purposes and for steam; indeed it is sometimes spoken of as steam coal, to distinguish it from the bituminous flaming coal of the Midlands and the North of

Head of Coal Mine Shaft

England. No wonder then that it is in universal demand and is distributed from the ports of Newport, Cardiff, and Swansea to all parts of the world.

Thus we see how it has come about that the coal-industry, which was at first no more than subsidiary to

[1] The upper measures are more bituminous but the farther westward the coal is worked the more anthracitic it becomes, even in the same seam.

The Works, Ebbw Vale

the working of the ironstone, has outstripped its fellow and become the largest and most important business in the whole of South Wales. Not that the manufacture of iron and its allied industries is dying out. On the contrary, as we showed in the last chapter, it is in a very flourishing state, but it has undergone this change, that whereas in former times the raw material was the native ironstone, it is now the ore imported from abroad, which contains a far larger percentage of the metal, and the native ironstone is now raised mainly for roadmaking and building purposes.

Other minerals worked or quarried in the county are clay, fireclay, sandstone, and limestone. It is the demand for the last mentioned that has produced the extensive quarries in the magnificent cliffs on the Wye above Chepstow, and destroyed their beauty for many generations. The stone is here shipped on the spot and conveyed down the river into the channel and so to Bristol.

As for the collieries, they are scattered over the valleys above Newport in every direction from Tredegar, Ebbw Vale, Beaufort, and Brynmawr in the north to Maesycymmer, Abercarn, and Risca in the south. In 1908 the output of coal in this district was 13,034,778 tons. Of this 3,931,572 tons were exported abroad, and 778,101 tons coastwise. The total amount of coal raised in the whole of the South Wales coalfield was 37,186,500 tons. Of ironstone only 11,263 tons were raised in the Newport district and 16 tons exported, while 512,346 tons of foreign ore were imported.

The following figures will show the comparative

Celynen Colliery, Abercarn

output of minerals in the year 1908 for the three mining districts of South Wales:

	NEWPORT	CARDIFF	SWANSEA
	Tons	Tons	Tons
Coal	13,034,778	24,708,294	12,484,041
Clay and Shale	4,335	134,044	7,100
Fireclay	94,235	56,419	31,807
Ironstone	11,263	8,093	1,998

Although so many tunnels have been everywhere driven into the hill sides, and so many coal mines sunk, little damage appears to have been done to the surface of the ground, except in the Rhymney valley, where cracks have appeared in the mountain and some subsidence has occurred.

14. Fisheries.

The Wye and the Usk have always been famous as salmon and trout rivers. Trout and grayling are taken above the tidal waters, salmon both in the tidal waters and above them; in the latter by means of rod and line, in the former by nets. There is no doubt that Wye salmon long ago found its way to the Bristol market, and when mail-coaches began to run it was also taken to Gloucester and London; nowadays it goes by rail, but any fishmonger on the river from Rhayader to Chepstow who applies for it may obtain a supply for the use of their local customers.

The fisheries were formerly managed by the lords of the manors to which they belonged, but they are now governed by boards of conservators, under the control of the Board of Agriculture and Fisheries. There are two of these boards concerned with the Monmouthshire fisheries, the Wye District Board, the headquarters of which are at Hereford, and the Usk and Ebbw Board, which has its headquarters at Newport. Each of them has jurisdiction over that part of the Channel which adjoins the mouths of their respective rivers.

On the Wye the line dividing the rod and the net fishing is at Brockweir. Below this point and including the Channel the nets in use are the draft net, the tuck net, the lave net, and the stop net. The draft net is 200 yards by 8 ; one end of it is held on shore, and the other payed out from a boat as far as it will go, when the boat immediately returns to land, and the net is hauled in. The tuck net is similar, but only 100 yards by 8, and is payed out between two boats, one at each end of the net. A lave net is a bag suspended from a wooden frame consisting of a handle and two moveable arms : it is held by a person standing in the water, who scoops up any fish that he can get. The stop net is also a bag net, but a temporary fixture. The fisherman moors his boat across the tide, and the net is stretched between two poles projecting over the side of the boat, in such a manner that the bag is carried under the boat by the stream : the man holds a string attached to the bag, and as soon as he feels a fish struggling in the net, he raises the poles and secures the fish.

There is another kind of trap, which however is not used actually in the Wye or the Usk, but only in the estuary of the Severn adjoining the mouths of these rivers. This is a long basket open at one end and closed at the other, called a " putcher." These putchers are never used singly, but are fixed in rows to a line of stakes, row above row, opening up stream. The salmon entering these baskets on the ebb tide are unable to turn round and get out again, and are therefore left high and dry at low water. The number of these nets and putchers in use is of course strictly regulated, and licences have to be taken out by all users of a rod, whether for catching salmon or trout.

In the Usk district the only net used is the stop net. Putchers are also used in the same manner as in the Wye district, but a distinction is here made between the putcher and the putt. The putt consists of three baskets dovetailed one into another, the last one being so close that nothing can get out of it. Thus while the putchers will only catch salmon, the putts will catch anything from salmon down to shrimps. Again, while the putchers are ranked in three or four tiers the putts are fixed in a single row.

The close season for salmon is as follows :—

	WYE	USK
Nets	16 Aug.—1 Feb.	1 Sept.—1 March
Rods	16 Oct.—1 Feb.	2 Nov.—1 March
Putchers and Putts	16 Aug.—16 April	1 Sept.—1 May

The close season for trout on the Wye is from October 2nd to February 14th, and on the Usk from September 2nd to February 14th.

As regards river pollution the absence of large towns in the basin of the Wye (the largest is Hereford) gives the river a tolerably clean bill of health, though some inconvenience has been felt in the upper reaches from reduction of the volume of water by the Birmingham waterworks on the Elan. With the tributaries of the Usk, however, the case is very different, and though the river itself above Caerleon is remarkably free from any defilement, the presence on their banks of the mining industries involving an influx of acid from the works, coal-dust from the collieries, as well as sewage from the towns, makes it impossible for any fish to live in the Afon Lwyd or the Ebbw and its affluents. And what has been said of these streams applies also to the Rhymney.

15. Shipping and Trade.

Chepstow at the mouth of the Wye and Newport at the mouth of the Usk are the only ports in our county. A century ago the trade of Chepstow was more than double that of Newport; at the present day Newport is the second port in the whole of South Wales, while the port of Chepstow is practically extinguished. The advantages of Chepstow were a river navigable for many miles inland, and the communication thus afforded with Hereford, Ross, Monmouth, and the Forest of Dean. Hence the trade in timber—ash, beech, elm, and oak— was considerable, beside that in various other commodities, among which may be mentioned the iron and wire sent from the works at Tintern. The Usk on the other hand is navigable only so far as the tide extends, and communicates with no place of importance.

But the march of events changed all this. First came the Monmouthshire Canal, which put the navigable portion of the Usk in connection with the mining industries of the hills, then the great development of those industries, and lastly the network of railways connecting every valley with the port. Another advantage was found in the river itself; as if to compensate for its short stretch of navigable water the Usk at its mouth is far deeper than the Wye, so that vessels which could only reach Chepstow at high water can be docked at Newport at any state of the tide. Thus the Wye ceased to be an artery of inland traffic, killed as it was by the railways and the greater facilities of Newport and other channel ports.

As the trade of Chepstow declined, that of Newport steadily increased. In 1842 the Town Dock covering 11½ acres was opened, and from this time the history of modern Newport may be said to begin. During the next half century trade continued to expand, and more dock accommodation on a larger scale, and suited for larger vessels, became an urgent need. At last in 1875 the Alexandra Dock, about a mile further down the river, was opened. This covers 28¾ acres and has an average depth of 30 feet. It is now known as the North Dock, and in 1893 it was connected with a new dock covering 68 acres, and called the South Dock. This in its turn is now in process of extension by the addition of another 75 acres, so that the whole of the Alexandra Docks when completed will form a floating harbour covering 171¾ acres. All these docks, including the Town Dock, are now owned and managed by the Alexandra Docks and Railway

Alexandra Dock, Newport

Company, which is the largest commercial undertaking in Newport. There are also dry docks for building and repairing vessels, as well as timber floats. One side of the docks is fitted with hydraulic coal-hoists and tips, and the other with moveable and fixed cranes, and warehouses for the general export and import trade.

A comprehensive idea of the trade of the port of Newport as it at present exists may be gathered from the following figures for the year 1908 :—

Total imports and exports   6,064,851 tons
Number of vessels cleared   4,028
Registered tonnage cleared   2,675,721

Of the exports coal is far greater than any other. In 1908, 3,931,572 tons were exported to foreign countries, and 778,101 tons coastwise. This does not include bunker coal, i.e. that consumed aboard ship. Then come "patent fuel" and iron and tin plates. Of the imports, iron ore, chiefly from Spain, heads the list with 512,346 tons ; then comes pitwood with 296,125 tons, and then iron and steel with 205,334 tons. Of course these figures do not include the coal and other products carried to other parts of the island by railway ; for example, more than a million tons of Newport coal are estimated to be conveyed to Cardiff and there shipped as Cardiff coal. Nevertheless Newport remains the third largest coal-exporting port in the United Kingdom, being surpassed only by Liverpool and Cardiff. No less than 29 ocean steam lines have vessels running from Newport to all parts of the world ; among the best known of these are the

A Train of Coal Trucks

Houlder Line, who have their principal offices in London and Newport, and the Clan Line, whose chief offices are in London and Glasgow.

The accompanying illustration shows the loaded coal-trucks on their way to the docks, but large as it seems, more than twice the quantity here shown is shipped daily. The process of transit from the mine to the ship's hold is

Coal Trucks on the way to Newport Docks

managed without a break, and the coal which in the morning was part of its seam in the depths of the earth is not seldom well on its way across the sea before the close of the day. The way in which the coal is transferred from the truck to the ship is most ingenious, and may be shortly described. Along the side of the dock may be seen several tall structures called coal-hoists, two of which are shown in the illustration, each approached by two roads, a lower- and a higher-level road. The lower-level is for the full trucks on their way to the hoist, and the higher-level for the returned empty trucks on their way from it. Each truck, it should be said, has a hinged flap at the end nearest the vessel, and is run from the lower or

Loading a Steamer from a " Coal-Hoist "
(Showing empty truck returning by high-level track)

" full road " on to the cradle at the bottom of the hoist. The cradle with the truck upon it is then raised by hydraulic power to the level of the top of the shoot, which extends from the hoist to the hatch of the vessel. The flap is now let down and the other end of the truck tipped up, thus discharging its burden down the shoot. The empty truck is then lowered to the stage where the higher or "empty road" leaves the hoist, and is at once run off to the sidings, whilst the next truck in waiting goes through the same process.

What is now the most conspicuous feature in modern Newport owes its origin to the rapid development of the port. This is the Transporter Bridge, with one

Trucks from Mine reaching the Surface

exception the only structure of its kind in the United Kingdom. It crosses the river about a mile and a half below the old stone bridge at a height of 177 feet above high water. It is supported at either end by lofty towers of open ironwork, and from the span is suspended the moveable car or platform on which the passengers travel. This car is on a level with the river banks, and is set in motion by electric power which winds the steel wire rope attached to the car round a drum fixed on the bank approached, at the same time unwinding the rope from the drum on the bank left behind.

The Transporter Bridge, Newport

16. History of the County.

We have already seen in the first chapter that when the Statute of 1535 called into existence the county of Monmouth, it united two distinct territorial divisions— Gwent to the east and Gwynllwg (Wenllwch) to the west of the Usk. Gwent was subdivided by the wooded range of hills stretching from Penallt to Caerleon, the western portion of which still bears the name of Wentwood. Gwent Uchaf or Upper Gwent lay to the north, and

Gwent Isaf or Lower Gwent to the south of this dividing line. As to the meaning of the term Gwent there has been much discussion, but the explanation best suited to the facts is that it signifies a clear open space, as distinct from one covered with wood. The lordship of Gwynllwg (Wenllwch) was acquired in the thirteenth century by the Clares, lords of Glamorgan, but it never became part of the latter county. From the Clares it passed to the Staffords, on the forfeiture of the last of whom, Edward Stafford, third duke of Buckingham, in 1521, it became the property of the Crown.

The written history of South Wales begins with the partial conquest of the Silures by Ostorius Scapula A.D. 50-51, and their final subjugation by Julius Frontinus, governor of Britain from A.D. 75 to 78. A military station was established at Caerleon (Isca), and as the Roman power became secure, a settlement was formed at Caerwent (Venta) some eight miles to the east. It is from excavations at these two places that our knowledge of the Roman rule in Monmouthshire is to be gained, for on the departure of the legions at the beginning of the fifth century, a dark cloud descends upon both of them, and when this lifts all traces of Roman civilisation are buried deep beneath the soil. Nor need we wonder at this result when we reflect that for six hundred years Gwent, and indeed the whole of Wales, was the scene of endless warfare, carried on either by the Welsh princes amongst themselves or against their English neighbours. These struggles need not detain us here; the age was one of destruction rather than construction, and even

the hill forts which are scattered far and wide over the
face of the country belong for the most part to a pre-
Roman time.   When at last at the end of the eleventh
century the Norman appeared upon the scene he began
to secure his position in fortresses of stone, the ruins of
which have lasted to our own day.

Caerwent: Winter Room of a Roman House with Hypocaust

The expedient adopted by the Norman king for the
subjugation of Wales was a peculiar one. Fully occupied
as he was in securing his conquests in England, and main-
taining order in his hereditary dominions across the sea, he
had no time to devote to carrying on a guerrilla warfare
against a stubborn race of mountaineers beyond the Severn.

He therefore gave permission to the most powerful among his followers to carve out and keep for themselves any lands in Wales which they could win at the point of the sword. Thus it was each man's interest to get all he could, while at the same time ambitious and possibly dangerous rivals were provided with a field for their activity. In this way, as we saw in the first chapter, were formed the Lordships Marcher, among whom were divided "the Marches" of Wales. March means a frontier or boundary, but in this case the boundary became so large that by the end of the thirteenth century it included the whole of what we now call Wales except the mountainous region north-west of the Dovey, and parts of Cardigan and Carmarthen. One of the first districts to be annexed by these adventurers would be Gwent, and Striguil (afterwards Chepstow) one of the first strongholds. This was founded by William Fitzosbern before 1070, and by 1094 the chain of Norman castles had reached to Pembroke, so that by this date the subjection of the low country to the south of the mountains may be regarded as complete.

Each Lord Marcher was a petty sovereign within the limits of his own territory: he had his own chancery and his own judges, and owned no superior but the king. His wars were waged independently either against the Welsh or against neighbouring lords. This state of things, it may easily be imagined, did not conduce to the establishment of order, and accordingly after an existence of some four centuries and a half it was abolished by Henry VIII and replaced by the rule of English law.

But even before this time the disorders and inconveni-
ences that arose from the quarrels between the various
lords, and the fact that there was one system of justice
in England and another in the Marches, had brought
about some kind of intervention on the part of the king.
In 1284 the Statute of Wales had settled the government
of Wales proper apart from the Marches, and the Princes
of Wales (of whom Edward of Carnarvon was the first)
and their council had been placed at the head of it.
There was therefore ready to hand an authority, the
powers of which were capable of extension, and accordingly in 1478 Edward IV established at Ludlow a Court
of the President and Council of Wales and the Marches
for the purpose of imposing the king's will upon turbulent
and refractory lords. The court was made permanent by
Henry VII, and became a kind of Star Chamber for
Wales. Its authority extended over Wales and the border
counties, and it continued to flourish throughout Tudor
times down to the outbreak of the Civil War, for the act
of union of 1535 left its powers untouched. At last after
the Revolution it was no longer needed and was finally
abolished in 1689.

Besides the Norman barons there gradually came
into existence landowners of another sort—the monastic
houses. Of these the principal ones in the district afterwards included in our county were the Benedictine priories
of Monmouth, Striguil, Usk, Bassaleg, and Abergavenny,
the Cistercian abbeys of Tintern, Grace Dieu, and
Lantarnam, and the Augustine priory of Llanthony.
All of them differed from the lay proprietors in the

fact that their estates were not liable to be subdivided, and that they were as a rule better landlords. Moreover their influence was uniformly exercised on the side of peace and order. At the dissolution royal favourites and adventurers, who cared more for rents than religion, entered upon their inheritance, and the splendid churches, if not deliberately destroyed, were left to fall to pieces stone by stone.

Apart from local disturbances, the shock of all the great quarrels which divided the nation down to the Tudor settlement was not unfelt in Gwent—the Barons' war, the struggle with Glyndwr, and the contention between the houses of York and Lancaster. But lying as it did on the fringe of all these momentous events, no great battle was fought within its limits. Glyndwr indeed visited his foes with fire and sword in Gwent as elsewhere, and the two occasions on which his men fought on Gwentian soil may deserve a passing notice. The first took place at Craig-y-Dorth, a spur of the Trellech range that overhangs the valley of the Trothy above the village of Michel Troy. Glyndwr had been badly beaten by Richard Beauchamp, Earl of Warwick, somewhere in what is now Montgomeryshire, but rallying after his reverse he had followed on the heels of his conqueror, and coming up with him at Craig-y-Dorth turned the tables on him and gained a signal victory. This was in 1404. On the second occasion, a year later, Glyndwr was not present himself, but his lieutenant Rhys Gethin attacked and burnt the town of Grosmont on the Monnow. The castle, however, was held by Henry Prince of

6—2

Wales with a strong force. Sallying out, the prince fell upon the Welshmen and utterly routed them, slaying 800 of them. In a foray of this kind it is doubtful whether there was much scope for the use of the long bow, for their skill in the use of which the men of Gwent and the neighbouring territories were so famous. It was with this weapon that they had fought under Edward I against Llewelyn at Snowdon, and had won the day at Crecy under Edward III.

At the outbreak of the great Civil War in 1642 Monmouthshire, like the rest of Wales with the single exception of Pembrokeshire, was held for the king. There were garrisons at Raglan, Monmouth, and Chepstow, all under the influence of Henry Somerset, fifth Earl and first Marquess of Worcester, the head of a family which was the most powerful in Monmouthshire from the days when it first became a county down to the last century. No fighting of importance took place in the open field, but Monmouth and Chepstow changed hands more than once, and only finally fell into the hands of the Parliament towards the end of 1645. Raglan on the other hand, where the Marquess himself resided, held out till August 1646, when it surrendered to Sir Thomas Fairfax. In the second Civil War, May 1648, Chepstow Castle was surprised by Sir Nicholas Kemeys, and held for the king for a fortnight. Then it was that Cromwell himself appeared before it on his way to Pembroke, but he was obliged to leave its reduction to one of his officers. After the Restoration Henry Marten, one of the regicide judges, was imprisoned within its walls till his death.

Under the Commonwealth Monmouthshire suffered severely for its Royalist sympathies. The power of the Somersets suffered an eclipse, and Chepstow together with a large slice of their estates in the south of the county was bestowed on Cromwell. The persecution of the church was perhaps more severe in our diocese than in any other part of the kingdom. As early as 1639 the first dissenting community in Wales was founded at Llanvaches on the road between Chepstow and Newport by William Wroth, the vicar of the parish, who had been suspended for disregarding the injunctions of his bishop; and a second was established shortly afterwards at Blaenau Gwent on the western mountains. Under the Puritan regime the ordinances against the established religion were prosecuted with the utmost rigour, and those whose duty it was to put them into force did not scruple to extend their legitimate powers. At the Restoration the tide turned in favour of the old order, but for the next eighty years the reaction against the tyranny of the Puritans went too far, and it was reserved for a great nonconformist preacher like Howel Harris to reawaken the sense of religion in the hearts of the people.

17. Antiquities—Prehistoric. Roman.

When we talk of the prehistoric period in the history of a country we mean the period before the commencement of written records. The date at which the historic period begins naturally differs in different nations. In

the south of England it may be fixed at the invasion of
Julius Caesar, 55–54 B.C., in South Wales at the struggle
of the Silures with Ostorius A.D. 50, and their subjection
by Julius Frontinus A.D. 75–78. But the absence of
written history does not mean that the prehistoric period
is utterly unknown to us. The men of those far remote
days have left traces of themselves and their work both in
the remains of their dwelling-places and in their cemeteries,
and to this day we find their weapons and their domestic
implements. These it is the business of the antiquary,
as distinct from the historian, to study and interpret. The
result of such investigations has led to the dividing of the
life of man on our island into three periods, based upon
the material used for the manufacture of weapons and
domestic implements : (1) the Stone Age ; (2) the Bronze
Age ; (3) the Iron Age ; but it must be remembered that
these terms are independent of dates, that the periods over-
lapped, and that they differed so much in different countries
that even at the present day we have some parts of the
world, e.g. New Guinea, which are still in the Stone Age.

In Europe, however, the Stone Age is divided into
two periods, the old or Palaeolithic, and the new or
Neolithic. The men of the palaeolithic age lived in holes
and caves, and their implements were roughly-chipped
flints. Our island was then connected with the continent
of Europe, and huge beasts such as the mammoth and the
woolly rhinoceros existed. Their bones, together with the
flint implements of palaeolithic man, are found in the deposits brought down by ancient rivers, and under the floors
of limestone caves. Such caves have been discovered on

the Great Doward and at Symonds Yat, both in Herefordshire, but close to the boundary of our county.

Whether, as many think, a long gap took place between
palaeolithic man and the coming of his neolithic successor,
or whether the two were directly successive in our country
is still a matter of controversy.   At any rate Britain had
become an island by the time that neolithic man appeared.
He was probably of Iberian race, the present form of
which is now represented by the Basques in the north
of Spain, while survivals are also found in the short
dark-haired inhabitants of South Wales.   The neolithic
implements are the earliest we find ground and polished,
though they exist in other forms and need not be further
described here. Neolithic man kept cattle, and grew crops.
He lived in huts hollowed out in the ground, and roofed
with turf or wattle and made vessels of baked clay
ornamented with a pattern.   On the approach of an
enemy he moved his family and cattle up to one of the
"camps" on the nearest eminence.   These camps are
very numerous on the hill-tops of our county, and were
afterwards often utilised and modified by the Celt.   The
neolithic men buried their dead in the long barrows which
may still be seen here and there on the hills.

After the Iberians came the Celts, an Aryan people,
who came over in successive divisions.   The first, called
the Goidels, drove neolithic man northwards and westwards before them and entered into their inheritance, only
to be themselves driven in the same direction by the second
horde of Celts, the Brythons or Britons.   The Goidels
were pushed north-westwards by the Britons to Scotland

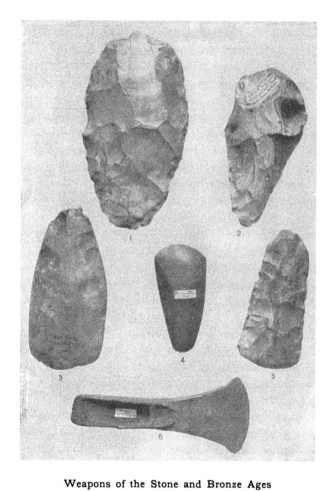

Weapons of the Stone and Bronze Ages

*1 and 2 are Palaeolithic; 3, 4 and 5 Neolithic;
6 is a Bronze Age palstave*

and Ireland, just as the Britons themselves when the Belgae came were pushed into Wales and Cornwall.

In prehistoric times, as we noticed in an earlier chapter, what is now the estuary of the Severn was a marshy plain, through which the river made its way to the sea. On our northern side this marsh extended right up to the hills, then, as is still partly the case, covered with the woodland which extended over the greater part of the county. Over marsh and forest roamed the great animals of which we have spoken, undisturbed till the earliest human settlers arrived and began to make their clearings and the tracks which led from one clearing to another. Long periods of time must have elapsed before the face of the country began to assume anything like its present appearance.

We have mentioned the Herefordshire caves just outside our boundary, but it was not till 1908 that any similar discoveries were made in our own county, when some quarrymen working at Ifton, not far from the Severn Tunnel, came upon some human skulls and other bones in a cleft of the rocks—the remains of a palaeolithic interment.

The long barrows and dolmens of neolithic man were an advance upon this primitive method. Of the former none have been noticed in Monmouthshire, but of the latter we have three examples, though the farmer and the road-maker are no doubt responsible for the destruction of many others. The finest is the Gaer Lwyd (or more correctly the Garn Lwyd), that is the Grey Cairn, the adjective grey perhaps having reference to its sepulchral character, just as the "Grey" Hill may have received

its name from the ghostly or *grey* associations of its burial
places. It is situated in the parish of Newchurch, close
beside the road from Chepstow to Usk made early in the
last century, and it was only the remonstrances of a local
antiquary which saved the stones of the dolmen from
being broken up for use in road-making. There are now
five supporting stones left, together with the huge capstone
12 ft. 5 in. long, 5 in. wide, and 9 in. to 12 in. thick, one
end of which has fallen, together with its support. In
spite of this the whole structure, which is 24 feet in
length, is still most impressive (see p. 174).

Another dolmen remains at Wern-y-cleppa near
Tredegar Park, and the fragments of a third at Heston
Brake near Portskewet. The Tredegar example has lost
its capstone, which has been broken into two pieces, one
piece only remaining on the spot, and it is altogether in
a more dilapidated condition than the Garn Lwyd. At
Heston Brake the enclosure lies east and west and is
divided into two chambers, the smaller one being towards
the west, but the covering stones are gone, and the side
stones, with the exception of two at the entrance on the
east, are not more than 3 ft. 6 in. high.

About two miles south-west of the Tredegar dolmen is
a large menhir or standing stone 10 ft. 6 in. high. Other
menhirs and various remains are found at Trellech and
on the Mynydd Lwyd (Grey Hill). The three stones at
Trellech are ancient boundary stones or hoarstones. The
two larger stones, the highest of which stands 14 ft. 4 in.
out of the ground, now lean considerably. On the
Mynydd Lwyd are three or four groups of burial mounds

consisting of earth and stones, some still containing cists, the most interesting of which is enclosed by a stone circle 32 ft. in diameter. There are also two menhirs 7 ft. 6 in. and 5 ft. 9 in. in height. All these remains have suffered much from treasure-seekers and others, but enough is left to show that this hill must long have been used as a cemetery by the early inhabitants of the district.

The prehistoric hill-forts are too numerous to mention. Almost every isolated height or projecting spur has been taken advantage of for the purposes of fortification. The one thing requisite seems to have been that the site chosen should command a tolerably extensive view over the lower ground at the foot of the hill, so that the approach of the hostile force might be descried. As an example we may take the Gaer Fawr, the largest in the county, on a projection of the Newchurch hills which commands a view northwards across the central portion of the county to the Black Mountains. It is about 300 yards in length and is protected by a double fosse. Sometimes one side of the fort is protected by a river, examples of which may be found below Piercefield, where the Wye forms the northern side of the entrenchment; and at Hardwick, where the precipitous cliffs on the eastern side fall directly into the river.

How the Romans came into Monmouthshire we cannot tell. They may have made their way southward from Gloucester and crossed the Wye above Chepstow, or they may have come across the Severn from Avonmouth to Portskewet. Or perhaps both routes were adopted. Be this as it may, the lines of the later Roman roads have been traced as follows: (1) the branch of Watling Street which runs southward from Wroxeter to Leintwardine, Hereford, Monmouth, and Caerwent, and thence crossing the Severn at Porthiscoed to Bath. (2) From Newnham to Lydney, Caerwent, Caerleon, Llandaff, and Carmarthen. (3) From Newnham to Monmouth, Usk, and Caerleon. (4) From Usk to Abergavenny and Brecon.

In the last chapter mention was made of the two Roman settlements at Caerleon (Isca) and Caerwent (Venta). Other Roman place names have been identified with Usk (Burrium), Abergavenny (Gobannium), and Monmouth (Blestium), but nearly all we know about the Romans in Monmouthshire is confined to the two places first mentioned. Both exhibit the ground plan of the four-sided Roman camp, and at Caerleon its angles are directed to the four points of the compass. The position is here defended on three sides by the windings of the river Usk, while to the north opens the valley of the Avon Lwyd. Considerable portions remain of the earthen rampart and stone wall which formed the external defences, particularly on the south-west side. On this side too, outside the walls, is the amphitheatre. The *principia* or headquarters of the legion (Legio II Augusta) was in the centre of the camp, where the church and churchyard now are. On the east side of the churchyard recent excavations have revealed the foundations of houses, and an exploration of the amphitheatre, which if completed promises results of the greatest interest, has been commenced.

But it is at Caerwent that the most extensive

excavations have been made. Here the population is smaller, and the modern houses comparatively few. The researches of the last ten years have resulted in the uncovering of about half the Roman town. Whether Caerwent was ever a purely military station is doubtful, but like Caerleon its ground plan is that of the camp,

Caerleon: Outer Wall of Roman Amphitheatre

and its four walls, which here face the points of the compass, are still standing. They are of immense strength and where best preserved are some 25 feet high, 11 feet thick at the bottom, and 8 feet at the top. The road from Chepstow to Newport, here proved by a transverse section to be Roman, enters the town at the east

gate and leaves it at the west gate. The sites of these
gates have long disappeared, but the north and south gates
have now been discovered, and in both cases the space
between the jambs had been blocked with masonry, leaving
only a narrow opening. It is conjectured that this was
done as an additional defence after the Roman troops had

Caerwent: Remains of Roman Temple

been withdrawn. Among the most interesting buildings
uncovered are the Basilica or Town Hall, with the
Forum in front of it, and a temple, the central chamber
of which has an apse at its northern end. The containing
wall of the amphitheatre has also been traced; the arena
it contains is much smaller than that at Caerleon, and the

tiers of seats instead of being built of stone were probably
of wood, like those in a modern circus. Moreover its
construction seems to have been an afterthought, for
foundations of buildings were found within the enclosure.
Another difference between the two is that at Caerleon,
where it is protected by the river, its situation is outside,
and here where there is no river or other external defence
inside the walls.

The reader may perhaps expect some mention of King
Arthur and his knights at Caerleon upon Usk, and of St
Tathan and his students at Caerwent, but we must refer
him for the former to the pages of that famous story-
teller Geoffrey Arthur, commonly called Geoffrey of
Monmouth, "the founder of the historical novel," and
for the latter to the *Lives of the Welsh Saints*[1].

18. Architecture—(a) Ecclesiastical.

The churches of Monmouthshire are of various types
and sizes and represent all the styles of Gothic architecture,
though the Perpendicular predominates. At their head
come the splendid abbey church of Tintern and the
priory church of Llanthony. Next to these we may put,
or rather might put, if only they stood as once they did,
the combined conventual and parish churches of Monmouth, Chepstow, Usk, and Abergavenny, but they are
all now mere fragments of their former glory. Among
the larger of the purely parish churches are St Woollos at

---

[1] Since this was written an interment of twelve skeletons, conjectured
to be those of St Tathan's twelve Canons, has been discovered at Caerwent.

Newport, Christchurch, Llantilio Crossenny, Grosmont, Trellech, Mathern, Caldicot, and Magor. All the larger and many of the smaller churches have belfry towers ; spires exist only at Monmouth, Trellech, Grosmont, Llantilio Crossenny, and Llanishen. In the others there is only a simple bellcot.

From the sixteenth century downwards, owing to a combination of causes, among which we may mention the relaxation of ecclesiastical authority, the growth of dissent, and the diminution of the rural population, the churches of the county fell into a bad state of repair. When, therefore, the middle of the last century brought with it the Gothic revival, they offered to the enterprising architects of that time an admirable field for the exercise of their ingenuity. In the case of the smaller churches—a large majority—the temptation to raze the existing building to the ground and replace it by an edifice in the approved style of nineteenth century monotony, too often proved irresistible. But in all the churches, small or large, with a few honourable exceptions, the craze for sweeping away everything that did not commend itself to the taste of the day was carried to an extreme, and left the present generation but too often an arid waste of scraped walls, encaustic tiles, and machine-made fittings. The ruins of Tintern and Llanthony have suffered much, but they have at least escaped this so-called restoration.

The small ruined churches of Runston and Sudbrook may be mentioned here. The former, situated on the hills above St Pierre, seems to have fallen into decay in the eighteenth century, when under the old poor-law small isolated villages were apt to become little more than nests of paupers and poachers.   It is a small Norman building consisting only of nave and chancel, an interesting specimen of its class—most of its fellows in the county having been restored out of all recognition—which even now deserves the preservation it does not get. Sudbrook now, owing to the encroachment of the sea, on the edge of the shore south of Portskewet, was probably disused earlier.   It was originally a church of the same type and still retains its Norman nave, but in the fourteenth century the original chancel was rebuilt and a south porch added to the nave.   Hence while the chancel arch at Runston is narrow and round headed, here it is wide and pointed.   Other ruined churches exist at Llangunnock, Llanfair Cilgoed, and Capel Newydd.

The church of Tintern Abbey still remains standing, but the northern arcade of the nave and the roof of the whole building have gone.   It was built in the latter half of the thirteenth century and is in the early Decorated style.   It includes the south side of the site of an earlier twelfth century church, traces of which may be seen in the north transept and in the north aisle of the nave.   This was the church of the original foundation in 1131 of Walter FitzRichard, lord of Striguil.   The conventual buildings, which in the absence of any reason to the contrary are usually on the south side of Cistercian churches, are here on the north side, which is the more sunny and open, and is easily accessible from the river.   On the south side the ground rises rapidly and

Tintern Abbey from the South-west

soon becomes a steep hill side. The remains of these buildings are extensive, but as compared with the church utterly ruinous. They are of various dates from the twelfth to the sixteenth century and are ranged round three sides of the cloister, the nave of the church occupying the fourth. The western range was the quarters of the conversi or lay brethren, the eastern that of the monks, while still further east was the infirmary, where the aged monks and the sick were accommodated. The whole area of some 27 acres which embraces the abbey and its precinct was surrounded by a wall, of which the south and west sides remain : on the north side is the river, and it is uncertain whether the wall was continued parallel with this or not. Both the outer and inner gatehouses which formed the entrance to Cistercian houses are gone, but the gatehouse chapel which stood between them forms part of a modern dwelling-house. The archway on the north leading to the ferry is still standing. The high road from Chepstow to Monmouth now runs through the precinct from end to end on the south side of the abbey.

The history of the abbey is uneventful. Founded, as has already been said, in 1131 by Walter FitzRichard it existed for just four hundred years and was dissolved by Henry VIII in 1535. Charters enriching the foundation with grants of land were bestowed upon it by several benefactors, but the most conspicuous gift and the one which appeals most to after generations is the new church built by a later lord of Striguil, Roger Bigod, 5th and last Earl of Norfolk of that family, who died in 1306. By

Tintern Abbey: the Nave, looking West

1287 enough was finished for mass to be said at the high altar. The north side of this church ran right through the original Norman church, which was pulled down as the new building advanced. The arms of Bigod appeared in the glass of the great east window. After the dissolution the buildings of the abbey were allowed to fall to ruin, and in a hundred years had assumed something of their present appearance. In the middle of the eighteenth century the church was cleared out and turfed, and though henceforward more carefully preserved, destruction was not fully stayed. At last in 1901 the abbey became Crown property; further foundations of the conventual buildings have been excavated, and the weak spots in the whole fabric have been made strong.

The Priory of Llanthony (Llanddewi yn nant Honddu) was founded for Augustine canons early in the twelfth century by the Norman Baron Hugh de Laci. Of the church then built, however, no traces remain, for the existing ruins are those of a church in the style of a hundred years later. The canons in fact became weary of their wild and unsheltered home, and during the disturbances of Stephen's reign they suffered so much from the raids of the Welshmen that under the patronage of Milo of Gloucester, Constable of England, and in 1140 Earl of Hereford, they migrated to Gloucester, where a new Llanthony was founded for them in 1136. From this time therefore there were two foundations, the old and the new. With the latter we are not now concerned; it continued to flourish like the parent church down to the dissolution, for the old foundation of the Honddu

was again inhabited when the troubles were over. The
transition Norman style of the present church leads to
the inference that the canons who returned found their
buildings in such a shattered state that reconstruction was
necessary. The arches of the nave are pointed, though
those of the triforium and clerestory are round headed.

Llanthony Abbey, from the North-east

At the west end, where the fully developed Early English
style appears, a splendid window of three lancets stood
down to the year 1803. Thus we see that both at
Tintern and at Llanthony the original Norman churches
gave way to the later ones. Unlike Tintern, however,
the ruins of Llanthony are still in a very neglected condition, and little is done to stem the tide of destruction

which has been proceeding for the last 450 years. The gatehouse (of a later date than the church) and part of the precinct wall are still standing. The gatehouse is now used as a barn.

Older than either of these ruined churches are the priory church of Chepstow and the church of St Woollos at Newport. Of the former, five out of the six bays of the nave, two stages of the west front, and the base of one of the piers that supported the central tower are all that is now left. The tower fell down at the beginning of the eighteenth century, and its materials were used for the western tower which replaced the gable of the west front. The massive square Norman piers of the nave supporting the triforium and clerestory, and the richly-moulded western doorway, show what an imposing Norman building once existed here. The priory was a cell to the Benedictine abbey of Cormeilles in Normandy. At the dissolution, when the church ceased to be used for monastic purposes, it is probable that its eastern end was destroyed together with the transepts, leaving the lantern and the portion used for parish purposes, namely the nave and aisles. With the fall of the tower the church was further shortened by the loss of the lantern, and the east end of the nave was then closed.

The nave of St Woollos—the anglican form of Gundleus, which again is the Latin form of Gwynllyw, the patron saint—is a fine example of the plain Norman. The aisles and chancel are Perpendicular and Decorated respectively, but the latter has been so pulled about in modern times as to be practically new. The Perpendicular

Chepstow Church: West Front

tower is at the west end, and between it and the Norman
nave is the peculiar feature of the church.   This consists
of an Early English galilee or antechapel, of a lower pitch
than the church itself.   At its eastern end it opens into
the nave by a very curious Norman arch supported by
two tapering columns of classical design with composite
capitals, part of which have been cut away to introduce
biblical designs.   It has been suggested that traditions
inherited from the neighbouring Roman station of Caerleon
may have influenced the artist who executed this work.
The tower has a niche on its west face containing a
headless statue, conjectured to be that of Jasper Tudor
Duke of Bedford (d. 1495) who built the north-west tower
at Llandaff Cathedral, and may have built the greater
part of this one.   He was lord of Gwynllwg (Wenllwch)
in right of his wife.

St Mary's Abergavenny was the church of the Benedictine Priory founded here in the reign of Henry I, but
there are now hardly any traces of Norman work, and
the present structure is mainly fourteenth century, though
the west front and nave were reconstructed in 1882.
The church is cruciform with a central tower.   The
nave has a north aisle, and the transepts are prolonged
eastwards to form chapels with side arches opening into
the choir.   The canopied stalls of the monks still remain,
but the glory of the church is the magnificent series
of altar tombs and effigies dating from the thirteenth to
the seventeenth century, and commemorating members
of the families of Cantilupe, Braose, Hastings, Herbert,
Powell, and Lewis.

Norman Doorway, St Woollos' Church, Newport

At Usk was a Priory of Benedictine nuns founded in the twelfth century. The church was cruciform with a central tower, but at the dissolution the choir and transepts were destroyed, leaving the handsome tower at the east end of the present church, its lowest story becoming the chancel. The tower arches are Norman, the nave with its north aisle fourteenth century. A strip of brass containing an inscription in Welsh was long a subject of dispute among antiquaries, but it has at last been satisfactorily interpreted and found to commemorate the chronicler Adam of Usk who died in 1350. The remains of the conventual buildings on the south of the church have been converted into a dwelling house.

The fate of the Priory church at Monmouth was that of the great church of Banbury in Oxfordshire half a century later. In 1737 the townsfolk decided that no attempt should be made to preserve it. They preferred to sweep away altogether the old fabric and traditions, and set up in its place a commodious edifice in the classic style of the day. With this their descendants had to be content till 1882, when it gave place to the present church, the only part of the eighteenth century work preserved being the tower and spire, which had then been rebuilt in the old style. The small Norman church of St Thomas the Martyr on the other side of the Monnow has been " restored."

Of the larger country churches we may mention Trellech, Llantilio Crossenny, Grosmont, Skenfrith, Mathern, Magor, Caldicot, and Christ Church. In the first three instances the church is larger than the present

population would seem to require, but the anomaly is
easily explained. Thus Trellech was formerly a town of
much greater importance than it is at present, and contains a mound which was probably a watching-place.
Llantilio would be the church of the garrison of White
Castle, and Grosmont, a corporate town, was burnt by
the partisans of Glyndwr in 1405, and never recovered
its prosperity.

Trellech and Llantilio resemble each other in having
fine Early English naves, the walls of which were raised and
clerestory windows inserted in the Perpendicular period,
but while the tower at Trellech is at the west end, that
of Llantilio is central. Both have spires, that of Trellech
being of stone, and that of Llantilio of wood with shingles.
Llantilio (the church of St Teilo) is said to be built on
the spot where in 560 A.D. a Welsh Prince, Iddon, aided
by the prayers of St Teilo, defeated the heathen Saxons.
In the fourteenth century the chancel and the adjoining
parts of the transepts were rebuilt in the Decorated style,
and a chapel, taking its name from Cil-llwch, an old
manor house in the parish, added on its north side. The
only remnant of the Norman church, which preceded the
present one, is the archway between the north aisle and
the transept, which was possibly the chancel arch.
Another noteworthy feature is the low arches which
support the tower.

The fine cruciform church of Grosmont is Early
English with Decorated additions. The octagonal tower
supports a broach spire. The proportions of the nave and
its lofty arcades are well seen, for it is not seated, the

choir and transepts being enough for the present size of
the parish.   Among the details a beautiful Early English
piscina with a trefoiled head beneath an arch containing
dog-tooth ornament should not be overlooked.   Skenfrith
is Early English and Perpendicular: the aisles do not
merely lean against the nave, but have their own roof-
trees and gables.   The low western tower terminates in

Rockfield Church and Lych gate

a wooden belfry, a common feature in the border country,
and found also in our county at Llanfihangel Ystern
Llewern, Wonastow, Rockfield, and St Maughan's.

Caldicot has a lofty tower with pyramidal roof
between the nave and chancel, and a large battlemented
south porch, the doorway of which has an elaborate
pinnacled and crocketed canopy.   Mathern in its style

and plan resembes Skenfrith, but it has a fine lofty
tower at the west end.   In the seventeenth century
a stone coffin was found in the church, believed to be
that of the patron saint, Tewdric, King or Prince of
Glamorgan, who fell in a battle against the Saxons.
Magor has a large porch like Caldicot, but on the north
side of the church; the doorway too is smaller and has the
window of an upper room above it.   The central tower
is Early English, the chancel mainly Decorated, and the
nave and aisles Perpendicular ; the latter are prolonged to
the eastern face of the tower, thus forming a substitute
for transepts.   A peculiar feature of the tower is a relic
of the days when the church was the fortress, as well as
the place of worship, of the parishioners : the parapet
projects upon corbels, thus enabling missiles to be hurled
down upon an assailant below.   These military towers
are also found at Shirenewton, Usk, Llangwm Uchaf,
Roggiett, Christ Church and elsewhere, while at Llan-
vetherine the whole of the upper part of the tower
projects over the lower.

Christ Church stands on the hill to the south of
Caerleon at the extremity of Wentwood.   It is in the
main a Perpendicular reconstruction of an earlier church.
Both nave and chancel have aisles with distinct roofs, and
the massive tower, the lower part of which is Early
English and the upper Perpendicular, is at the west end
of the south aisle.   Its south side however is not in a line
with the aisle wall, but projects beyond it, a fact which
seems to show that if the aisle replaced the nave of an
earlier church it was laid down on a narrower plan.

Most of the small towerless churches have been rebuilt, but some interesting Norman features have been preserved at Malpas. Among the smaller churches with towers which still preserve their continuity with the past may be mentioned Llangibby and Llanover. The latter still has the exterior of its nave and chancel whitewashed, the traditional practice of the Welsh when the building was of rubble. Its chancel, too, is a step lower than the nave, a peculiarity seen also in Mamhilad, and, before modern alterations, in Wolvesnewton.

Rood screens and lofts have almost all disappeared, but splendid examples, the work of local craftsmen, survive at Llangwm Uchaf and at Bettws Newydd.

19. Architecture—(b) Military.

No county for its size has more ruined castles than Monmouthshire. Not that the Welsh were castle-builders, they were content with the entrenchments noticed in an earlier chapter; it was the Norman invaders who were the first to fortify themselves behind walls of stone, and even the Normans on their first incursions had to put up something less durable. Those artificial mounds, sometimes from 50 to 60 feet high, which are common enough all over the country, and of which we have examples at Caerleon and at Caldicot, are now believed to be of Norman construction. The first thing the Norman adventurer had to do, when once he had gained a footing on Welsh soil, was to erect a stronghold from which he might command the surrounding district and terrorise its inhabitants. He therefore began by throwing up one of these mounds, unless indeed he happened to find a natural mound ready to his hand, and surrounded the top of it with a stockade. In the enclosed space would be a wooden building which served as his temporary quarters. We thus get a palisaded mound surrounded by a ditch, from which the earth for the mound had been taken. The next step was to enclose and strengthen with a palisaded entrenchment a court or "bailey" across the ditch and at one end of the mound: this would contain wooden shelters for the soldiers and stables for the horses. But the Norman lord was not long satisfied with such an elementary style of fortification, and in a year or two he began to find himself sufficiently established to build something more solid. This was the stone castle, of which we now have to speak, and of which, as may still be traced in the earliest examples, the mound and bailey formed the nucleus.

The strongest part of these castles was called the keep; in this the garrison might take refuge after the rest of the defences had been taken, and hold it for an indefinite period. The keep was of two kinds—the shell keep and the rectangular keep, of both of which we have examples in our county. The shell keep took the place of the stockade on the summit of the mound; it was many sided and followed the shape of the mound on which it stood: the enclosure formed an open court, against the sides of which the various sheds and dwellings were built. A keep of this kind exists at Abergavenny and once existed at Caerleon. The rectangular keep was a massive stone tower, consisting generally of a basement and two upper stories. If the mound was a newly-made one and not strong enough to support its weight, it was erected in the bailey. We have a rectangular keep at Chepstow, but more typical examples may be seen elsewhere as at Kenilworth, Rochester, and the Tower.

These keeps, the shell and the rectangular, continued to be built down to about the end of the reign of Henry II, when they were abandoned in favour of the round tower or cylindrical keep, of which Pembroke is one of the best-known instances, and of which we have a fine example at Skenfrith, where it stands in the centre of the enclosure. From their use as keeps these round towers were soon employed for the purpose of strengthening the angles of the outer walls of the castle, as well as the intervening curtain walls. In this way we find them used in our thirteenth century castles as at Grosmont, White Castle, and Llanfair. Two of them, as at White Castle, often form the gatehouse.

In the fourteenth century another style of castle-building was introduced, called the concentric, because the typical form is arranged in lines of defence one within the other, so that when the outer line had been taken by the enemy the inner could still be held against him. Thus the innermost court or ward served the same purpose as the keep in the Norman castle, for curtain walls with their elaborate parapets and projecting bastions were sufficiently strong of themselves. At the larger castles, as

at Caerphilly, there were three of these concentric enclosures, at the smaller, as at Beaumaris and Harlech, two.
In our county we have no typical example of the fourteenth century castle. Additions in this period were
made to some of our older castles, as we shall see directly,
but it must be noticed that where the site of the castle
was a confined space naturally or artificially defined, as at
Chepstow, there was no room for these wards to be
built one within the other ; they had so to speak to be
telescoped out, so that the wards covered the whole available space, one behind the other. Marked features of
these later castles were the hall and the chapel, which
had formerly been confined within the walls of the keep.
They were now built on a larger scale, and usually
attached to the interior face of the curtain of the inner
ward.

Early in the twelfth century the frontiers of what
was destined later to become our county were well
guarded on both the east and the west sides. On the
Wye were the castles of Monmouth and Striguil (Chepstow) and on the Usk those of Abergavenny, Usk,
Caerleon, and Newport : later came Caldicot in the south,
and Grosmont, White Castle, and Skenfrith in the north,
while the tenants of the lord of Striguil guarded the
forest of Wentwood by the smaller fortresses of Dinham,
Llanfair, Penhow, Pencoed, and the Edwardian fortalice
called Cas-Troggy.

As the country became more settled most of these
castles were allowed to go to ruin, those which were kept
up being either those of first importance and commanding

strategic positions, such as those of Monmouth, Newport, and Chepstow, or those still used as the family seat of their owners, such as Pencoed and Penhow, which will be considered in the next chapter.

Let us look at Chepstow, one of the finest feudal castles remaining in Britain. From the Gloucestershire bank of the river, from which its whole length of some

Seventeenth Century House, Monmouth Castle

250 yards may be surveyed, it appears to be an upward continuation of the steep limestone cliffs which here rise from the water's edge. On the other side towards the town it overhangs a deep ravine out of which some of the stone used in its construction was quarried, but much of it is oolite and came from the quarries belonging to the Clares on the edge of the Cotswolds between Cheltenham

Chepstow Castle

and Painswick.   The building of the original castle by William FitzOsbern, Earl of Hereford and of Breteuil, is recorded in Domesday.   This was probably soon after the Conquest and at any rate before 1072, the date of his death.   The site thus chosen by FitzOsbern was an important one, commanding as it did the passage into those fertile districts of South Wales between the mountains and the sea, which at that time crossed the Wye by a bridge about half a mile above the present one, where the Roman road from Gloucester to Caerwent comes down from Tutshill and ascends a steep gully on the other side. It was no doubt on the rock at the highest and westernmost point of the present castle, now forming the fourth court, and well within sight of the passage over the river, that the first Norman stronghold was established, and protected on its eastern side by a cross ditch cut through the rock at right angles to the river.   The subsequent history of the castle is that of its gradual extension eastwards from this point along the platform of rock which lies between the river and the ravine.

In the absence of contemporary records any attempt to fix the dates at which the additions to FitzOsbern's work were made must be conjectural, but the process would seem to be as follows : in the twelfth century the third court, containing the massive rectangular keep, and the second court again to the east of the third were built along the platform eastwards, and in the thirteenth century the first court was built to the east of the second. It must not however be supposed that all these buildings remain exactly as they were when first erected. Subsequent alterations and minor additions have modified them considerably, and particularly is this the case with the great Norman keep, the first floor of which over the basement or cellar was divided transversely about the end of the thirteenth century into a hall to the east and a solar or parlour of smaller size to the west.

Finally, on the completion of the castle, the main

Newport Castle

entrance was transferred from the western extremity to the eastern, and the first court became the most inhabited portion.   On the north side of this court is the hall with kitchen, pantry, buttery, cellars, and other apartments. The great tower at its south-eastern corner, in which Henry Marten was confined after the Restoration, was occupied by a basement and three spacious rooms one

above the other, while to the west of it against the south
wall are traces of other buildings. It is worth noticing,
in illustration of what has been already said, that with
the exception of the south-western tower, all the towers
of the older part of the castle are rectangular, while those
in the later parts are round.

Of Monmouth, Abergavenny, and Newport castles,

Caldicot Castle

(*Before the recent alterations*)

the remains are small ; at Caldicot, now partially restored
and inhabited, we have the shell, with a very splendid
rectangular gatehouse, now repaired, on the south side
in the transition Norman style ; the large round tower
which stands on a moated mound in the north-western
angle of the enclosure is probably the keep of the original
fortress. White Castle, so called because once covered

with a coating of white plaster, stands on a hill about five miles south of the Monnow, and within easy distance of the castles of Grosmont and Skenfrith. It is a good example of the late Norman type, and is a strong plain building undisturbed by later additions. Together with the castles just mentioned it effectually secured the north of the county, and was supported on the west and east by the castles of Abergavenny and Monmouth. There is no keep, but the curtain wall has the shape of an elongated hexagon defended by six lofty round towers, the two on the north forming the gatehouse. The castle itself is surrounded by a deep moat, outside which on the north and east is a large space enclosed by an outer moat. The northern part of this space is walled in except on the side facing the castle, and strengthened by four mural towers, thus forming a kind of barbican or outwork. On the south side is another and very much smaller outwork in the shape of a half-moon with the concave side towards the castle; this is also defended by a moat which on its convex side is really an extension of the main castle moat. From the above description it will now be clear that White Castle belongs to the style intermediate between the early Norman and the Edwardian or concentric. In the fifteenth century it became ruinous, but the great strength of the masonry has so far preserved it from utter decay.

Grosmont and Skenfrith probably became ruinous about the same time. The former is surrounded by formidable earthworks and situated on the cliff above the river, the latter on the other hand lies close to the water, and owed its protection to the marshy character of the ground. Its remains consist of a circular keep crowning a low mound, and situated in an area enclosed by four curtain walls with projecting round towers at the angles. There was also a third, or counting Monmouth a fourth Monnow fortress about five miles above Grosmont at Oldcastle, but of this nothing but a few entrenchments remains. There were also castles at Castell Glâs or Greenfield Castle at the mouth of the Ebwy and Roger-stone (called in Welsh Tre-Gwilym), near Bassaleg. Raglan will be dealt with in the next chapter.

20. Architecture—(c) Domestic.

The style of house-building characteristic of any particular district was formerly dependent upon the materials most easily accessible. In the east and south-east of England it was brick, in the oolitic region of the midlands a durable freestone, in the chalk counties flint relieved by brick, and along the western border from Lancashire to Worcestershire and Herefordshire a com-bination of timber and plaster. The builders were the local masons and carpenters, and not only was their work always sound, but they were often capable of producing the most beautiful effects. Each in his own fashion made the best of such material as lay ready to his hand ; in a stone country he did not go out of his way to use brick, in a clay country he had no means of obtaining stone. There was therefore a marked individuality about the domestic architecture of each part of the country, which improved means of communication have too often replaced by a depressing monotony. With the arrival at the nearest railway station of trucksful of bricks and slates, together with wagonloads of ready-made doors, window-frames, joists and other fittings, the traditional style has

Horse and Jockey Inn, Llanfihangel Pont-y-Moel

been driven out, and nowadays the bricks and mortar are put together in much the same fashion in Cornwall as in Cumberland.

In South Wales brick is an alien, the native material being the local stone. The farmhouses, cottages, and

even churches built of this stone were always white-
or yellow-washed, and roofed with stone or thatch.
The limewash keeps out the wet, and the houses thus
treated harmonise with the landscape and have a not
unpleasing effect.

As for the towns, in the poorer streets the whitewashed cottages are still found ; in the better streets the

Raglan Castle

houses are covered with "rough-cast" and yellow-washed,
while the fronts are often painted.   In quiet old towns
such as Usk and Chepstow many houses of this kind
remain.   Newport is of course of more modern type.

Of the larger country houses much the finest was
the Marquess of Worcester's princely mansion at Raglan,
the ruins of which are the property of his descendant

the Duke of Beaufort. Its surrender to Fairfax in August, 1646, has been mentioned in an earlier chapter, and its demolition followed in accordance with the customary policy of the victors. The buildings are arranged round the sides of two large courts, the curtain walls of which include towers with salient angles towards the exterior or field. Between the courts stand the hall and chapel side by side ; the walls of the former with its great oriel at the dais end are entire, but the whole of one side of the latter is gone. To the north, looking towards the Black Mountains, are the apartments occupied by the King on his two visits to Raglan in 1645. On the south side of the castle is a detached hexagonal tower of great strength surrounded by its own moat, known as the Yellow Tower of Gwent, perhaps from the yellow-lichened surface of the stone. It is a magnificent piece of masonry, but two of its six sides were destroyed at the time of the demolition. It was probably built towards the end of the fifteenth century, before the main body of the castle, which dates from the sixteenth and seventeenth centuries. The unrivalled splendour of Raglan was a fitting appanage to the predominance of the Somersets over all the other families of the county. Their founder, who rose to distinction under Henry VIII, strengthened his influence in these parts by a Welsh alliance, and his descendants became the largest land-owners in the county.

We may now pass to the dwellings of the lesser gentry, of which we shall find few traces of an earlier date than 1485, when, with the accession of a Welsh

Raglan Castle: the Yellow Tower of Gwent

sovereign, border conflicts came to an end, and his countrymen could turn their attention to the arts of peace. Before this time the better class of house was generally built of timber, which was then more abundant, and if situated in a valley was protected by a moat. Accordingly, when these timber houses began to give place to stone, the new house was constructed on the same general plan as its predecessor.

The principal apartment was the hall, which was separated from the kitchen by a passage running right through the house with an outer door at either end of it. At the lower end of the hall next the passage was a wooden screen in which were doors opening into the passage, and at its upper end was the dais with its "high table." The arrangement, in fact, was the same as that which obtains in our Universities and Inns of Court to-day. Sometimes a withdrawing room was added later at the upper end of the hall, and it was not long before both hall and kitchen were covered by an upper storey. As the demand grew for more accommodation the original plan was doubled by the addition of a second house as it were to the side of the first, an arrangement which we find at Treowen, a fine Jacobean house about three miles south-west of Monmouth. The date of the house is furnished by the screen, which bears the inscription "W. I. 1627," the initials being those of William Jones, the then lord of the manor.

At Penhow, about halfway between Chepstow and Newport, is the original settlement of the Seymour (St Maur) family. The most ancient part of the house

is a rectangular tower of the twelfth or early thirteenth
century with a projecting parapet resembling those of
the church towers already noticed, and reminding one
of the peel towers of Northumberland.   To the south
side of this tower was added a forebuilding or entrance
porch.   An enclosed court no doubt rendered the original
dwelling complete.   The next additions were made in the
fifteenth century, when a hall and other apartments were
built along the south side of the court, and lastly in the
seventeenth century, probably soon after the Restoration,
the north-west angle of the court was filled up with a
dwelling-house, one end of which abutted on the north
side of the tower.

A mile and a half south-west of Penhow is Pencoed,
in the thirteenth century a moated house belonging to a
family called de Mora or de la More.   Very little of the
original house beyond a round tower at the south-west
angle of the court is left, for in the time of Henry VII
or Henry VIII it was rebuilt by a branch of the Morgan
family into whose possession the estate had then passed.
The remains now consist of a ruined gatehouse in the
centre of the west side of the court, and opposite to it a
range of buildings, now used as a farmhouse, containing
the hall, kitchen, and other offices, together with the
upper floors, which include a gallery over the hall.   But
the northern part of this range is now a ruin, and the
remaining part has been altered by modern partitions.
Two noteworthy features, however, are still conspicuous,
a handsome three-storied porch on the west or court side,
through which the hall is entered, and a strong tower

containing the staircase, projecting from the east or exterior side of the house.

In the south-east part of the county also are the mansions of St Pierre and Itton (which retain their gatehouses of fourteenth or fifteenth century date), Mathern Palace, and Moyne's Court. Mathern Palace was for three centuries (1406-1706) the residence of the

Bishop's Palace, Mathern

bishops of Llandaff, whose palace at Llandaff was burnt by Glyndwr. The gatehouse by which the court was entered was destroyed in the eighteenth century, but the house itself, which dates from the fifteenth century, remains, and after having been used as a farmhouse for nearly two hundred years has at last fallen into good hands, and is cared for as it deserves. Moyne's or

Moine's Court, which takes its name from a family called Moigne, who once owned it, is a fine, many-gabled house not far off, of late sixteenth or early seventeenth century date. The turreted gatehouse remains, but if there were ever any buildings on the right and left sides of the court they have disappeared.

In the same neighbourhood is the Ty Mawr or "Great House" of Crick. It is now a mere shell, and

Moyne's Court

a modern house has been joined on to one end of it. It seems to have been built in the fifteenth century to take the place of an earlier moated house, the site of which can still be traced hard by, and to have received alterations in the time of Elizabeth. In the farmyard is a chapel, apparently built in the thirteenth century, and now used as a barn. As the chapelry of Crick was founded much earlier than this it probably succeeded a timber structure. It was to this house that Cnarles I rode over from Raglan in July 1645, in order to meet Prince Rupert, who crossed over from Bristol for the purpose.

On the western side of the Usk are the old manor houses of Pentrebach, Tredunnock, and St Julian's. The first was a grange belonging to the neighbouring abbey of Lantarnam, and the earliest part of the present house appears to have been built about the time of the dissolution: it is of the orthodox type with cross passage through the house dividing the hall with its oak screen from the kitchen. About a century later a second wing of the same character was added at right angles to the old one, in such a way that the side of the new hall was built against the end of the old kitchen: the remarkable feature of this new building being that it was of brick and not of stone. The Ty Mawr at Tredunnock was another interesting old house of the same type, but its original plan has been altogether concealed by modern alterations and additions. The ancient mansion of St Julian's has been little altered externally. Its chief interest lies in its association with Edward Herbert, first Lord Herbert of Cherbury (1581–1648), who acquired the estate by marriage with the heiress, and occasionally resided there.

Other interesting old manor houses are, or were, Cil-llwch near Llantilio Crossenny, The Waen at Skenfrith, Perthir near Monmouth, Wernddu near Abergavenny, Crindau near Newport, Kemeys Inferior, Machen Place, Bedwellty Place, and Penllwyn-Sarph on the Sorwy.

Among the larger houses still inhabited we may also mention Llanfihangel Crucorney Court, which dates from the sixteenth century; a group of seventeenth century mansions of the Inigo Jones type including the Castle House at Monmouth, the front of Troy (i.e. Trothy)

Troy House

House near the confluence of the Trothy with the Wye (the back is early Tudor), Llangibby, Pontypool Park, Llanarth, and Tredegar; and the eighteenth century houses of Llanwern and Piercefield; while the nineteenth century may be represented by the Hendre.

21. Communications—Past and Present.

By the communications of a country we mean the methods employed for getting from one part of its surface to another. These have been hitherto limited to three, road, rail, and water; a fourth method, the air, promises to add itself to the others, but in its present elementary stage we can afford to disregard it.

The lines of the Roman roads that pass through our county have been indicated in a former chapter: it is only here and there that they coincide with our modern highways. This divergence may be partly explained by the fact that the Roman road-makers were not so careful to avoid obstacles as those of modern times; not that they did not often choose to go round a hill when practicable rather than over its summit, but in a very mountainous country like Wales and the borders, they generally marked out a pretty straight course irrespective of the inequalities of the ground. Partly also it is due to the fact that the river valleys in the absence of drainage and cultivation were in early times mere swamps, overgrown with alders, willows, and coarse herbage. Indeed, it is only in quite modern times that the roads, and following them the railways, have taken to following the winding courses of the rivers; what is lost in distance being gained in speed.

The Romans were magnificent engineers, as may be seen by a section cut through any of their roads. When the course had been marked out, the ground was levelled and beaten down till a hard foundation was formed ; on this were placed three layers of small stones, lime, sand and other materials, while the surface was formed either of cut stones or, more frequently, of a firm

Old Roman Road, Mamhilad

bed of gravel and lime. When they left the country there were no roads to be compared to theirs till the days of Telford (1757-1834) and McAdam (1756-1836). In South Wales there were no roads at all that we should now consider worthy of the name down to the end of the eighteenth century. "We travel in ditches," said a gentleman of Monmouthshire at the bar of the House of Commons when giving evidence in favour of the first Turnpike Act. In many places these hollow ways or "ditches" may still be seen by the side of the modern road. According to statute all sorts of the common people were compelled to labour for six days in the summer on the repair of the roads, and this labour was supposed to be overlooked by two surveyors elected in every parish, but this obligation was easily shirked, and a writer of Elizabeth's time complains that scarcely two good days' work out of the six were well performed. We may well believe that if big stones were thrust into the largest holes, and a few loose stones from the adjoining fields were scattered over the top, these were all the repairs that our Monmouthshire roads ever had. In such roads, of course, there could be no wheel traffic. People travelled on horseback, and goods were transported on long trains of pack-horses or mules.

The load carried by a horse was usually three hundredweight, or four bushels of grain, and on the old roads two and a half miles an hour was considered an expeditious rate of progress. When the farmers began to use wagons they had to send on in front of them a boy blowing a large horn to give notice to any other vehicle approaching to halt at some place where the two could pass each other. If this warning was disregarded, horses had to be attached to the tail of one of the wagons to drag it back till a passing-place could be found. Besides the horn, a hoop was fastened to the collar of the fore horse from which were suspended half-a-dozen large bells, which kept up a perpetual jangling as long as the horse was in motion. As for travellers between London and South Wales who did not ride on horseback, the only means of conveyance was the stage wagon. One of these left Monmouth about two on Monday morning arriving in London on the following Saturday evening, while the return journey was made in the following week, so that by this method of travelling it was possible to leave Monmouth for London only once a fortnight. Private carriages were very rare in the county before the turnpikes. The first country squire to visit Monmouth in his own carriage was Capel Hanbury of Pontypool, when sheriff. He was attended by a troop of labourers to open gates and pull down hedges where the road was too narrow, and it took him nine hours to drive the twenty-one miles.

At last in 1756 came the first Turnpike Act, but for some time it effected but little improvement, for Arthur Young, who travelled from Chepstow to Cardiff in 1767, writes as follows :—"But, my dear Sir, what am I to say of the roads in this county ! The turnpikes ! as they have the assurance to call them, and the hardiness to make one pay for. From *Chepstow* to the halfway house between *Newport* and *Cardiff* they continue mere rocky lanes, full of hugeous stones as big as one's horse, and abominable holes. The first six miles from *Newport* they were so detestable, and without either direction-posts, or mile-stones, that I could not well persuade

myself I was on the turnpike, but had mistook the road."
By later Acts passed in 1799 and 1810 the roads were
improved and made passable for stage-coaches. The
payment alluded to by Young was the toll which was
collected from travellers at the turnpike gates which
barred the road : the gates have long been gone, but the
toll houses often remain. The gates of a fixed district
were let to the highest bidder by a "Turnpike Trust,"
which also had the duty of keeping the roads in repair.
These roads are now managed by the County Council,
and the parish or cross roads by the District Council.

In the seventeenth century the best known roads
in Monmouthshire were those from Gloucester by
Micheldean, Monmouth, and Abergavenny to Brecon;
from Monmouth by Trellech (whence a branch led to
Chepstow), Devauden, Wentwood, Cat's Ash, and
Newport to Cardiff; and from Micheldean through the
Forest to Chepstow. As enabling us to form a picture of
the country at the time it is interesting to note that the
road from Monmouth to Brecon ran through an enclosed
country all the way, while that from Monmouth to
Cardiff ran through open country except for a short
distance near Monmouth. The road from Trellech to
Chepstow, which went near Trellech Grange and by
Tintern Cross, leaving the Abbey about a couple of miles
to the left, was almost wholly enclosed.

The principal bridges (the lesser streams were forded)
were those over the Wye and over the Monnow at
Monmouth, over the Wye at Chepstow, and over the
Usk at Caerleon and at Newport. At Monmouth,

where a rapid rise in the rivers could only be caused by a flood, both bridges were of stone. At Chepstow, Caerleon, and Newport, where apart from any disturbance caused by floods in the higher reaches, the level of the water was constantly increasing or decreasing owing to the flow and ebb of the tide, the bridges had to be more lofty and were therefore built of wood.

Bridge over the Monnow at Monmouth

All three bridges were built on the same plan, which reminded Archdeacon Coxe of Caesar's bridge across the Rhine. The roadway was perfectly level, and was carried on wooden piers resting on platforms of stone; each pier consisted of three uprights braced together by cross-pieces, the whole standing edgeways and being strengthened at either side by a slanting beam or

buttress facing up and down stream. These wooden
bridges have long disappeared. At Newport and Caerleon
their place has been taken by massive stone bridges,
the former erected at the end of the eighteenth century,
and the latter somewhat later. At Chepstow the existing
iron bridge, the second ever built of that material, was
put up in 1815–16. As already stated in Chapter 5 the
ordinary rise and fall of the tide at Chepstow is about
45 feet, but exceptionally high tides have been known to
reach 48, or even 49 feet above low water mark. Under
such circumstances it is clear that a wooden bridge would
be much more easily engineered than a stone one, as well
as more easily repaired when damaged. The two counties
of Gloucester and Monmouth meet in mid-stream, and
each county keeps its own half of the bridge in repair.

During the first half of the last century the mail
coach routes were from Gloucester to Ross[1], Monmouth,
Raglan, and Abergavenny to Brecon; from Bristol *viâ*
the New Passage to Black Rock, Newport, Bassaleg and
Cardiff; and from Chipping Sodbury to Aust and across
the Old Passage to Chepstow. There were of course
other coach roads connecting the chief towns, and
numerous cross roads.

There was formerly a considerable river traffic
between Monmouth and Bristol. Goods were taken
down the Wye as far as Brockweir in barges, and there
transferred to "trows," wide flat-bottomed schooner-
rigged vessels, which did the rest of the journey. There

[1] The road by Ross began to rival the old road *viâ* Micheldean and
Coleford about 1780.

was also a regular trade between Chepstow and Bristol by the same route. The cross-channel trade between Bristol and Newport still flourishes, and some lighter craft ascend the Usk as far as Caerleon.

Canals, which came into vogue in the last quarter of the eighteenth century, are confined to the west of the county. Great expectations were entertained of their

The Brecon and Newport Canal, near Newport

utility, and were realised until the advent of railways. The Monmouthshire canal was begun in 1792 for the purpose of transporting the mineral traffic of the hills to Newport. About a mile above that town it divides into an eastern and a western branch. The western branch ascends the Ebbw valley by Risca and Abercarn to Crumlin, where it stops; while the eastern branch follows a fairly straight course northwards past Malpas and Pontnewydd to Pontypool. Here it originally came to an end, but a few years later it was continued under the name of the Breconshire Canal across the Afon Lwyd and round the eastern and northern sides of the Blorenge into the Usk valley, and so to Brecon. This continuation took several years to com-

The Lock, Alteryn, near Newport

plete, and it was not till 1812 that the towns of Brecon and Newport were connected by a complete system of inland navigation.

A glance at the map will show that the western valleys, or the hundred of Gwynllwg (Wenllwch), have the appearance of a network of railways. Every valley has its line, while two other lines cross the hills from east to west.

The iron road was not here the novelty which it was in the agricultural districts. The revival of the iron industry about the year 1760 called for a more efficacious means of transit than trains of mules or pack-horses. Railroads were accordingly constructed on the same principle as those of our own time for the conveyance of wagons or "trams" laden with minerals. The point of difference was that the traction was by horse and not steam power, or in the case of an incline by the sheer weight of the trams themselves.

The first of the modern railway companies to turn their attention to Monmouthshire was the South Wales company since absorbed in the Great Western. Their line from Chepstow to Swansea was completed by 1850. Meanwhile Brunel was building his monster bridge across the Wye, by means of which the line was continued to Gloucester. This bridge was ready for traffic in 1853. It is too well known to need description; suffice it to say that its distinguishing feature consists in two great overhead tubes which help to support and steady the permanent way.

In the same year, 1853, the Newport and Hereford company opened their line from Newport to Abergavenny and Hereford. One of its Herefordshire stations "Tram Inn" commemorates an earlier tram line. This line has also been absorbed by the Great Western. The only other lines in the county to the east of the one last mentioned are the line from Pontypool Road to Monmouth, which is continued along the Wye to Ross; the Wye valley line, opened in 1873, from Chepstow to

Monmouth; and the Severn Tunnel line to Bristol. This last route formerly crossed the water by a steam ferry, but in July 1887 the tunnel took its place. This great tunnel, 4 miles 624 yards long, of which $2\frac{1}{2}$ miles are under water at high tide, took nearly 14 years to finish, the works having been more than once flooded by the tapping of springs of water.

The labyrinth of the Gwynllwg (Wenllwch) lines we

Crumlin Viaduct

shall best understand if we take (1) the two east and west lines, (2) the six north and south. (1) The cross line to the north is the L. and N. W. R. line from Abergavenny to Brynmawr and Merthyr Tydfil: this has a short branch from Beaufort to Ebbw Vale. The cross line to the south is the G. W. R. line from Pontypool to Aberdare and Neath. It is this railway which crosses the Ebbw valley by the Crumlin viaduct at a height of 210 feet. (2) Taking the north and south lines from east to west we have (i) the G. W. R. line from Pontypool to Blaenavon; (ii) the L. and N. W. R. line from Pontypool to Brynmawr: these two lines follow the Afon Lwyd; (iii) the G. W. R. line from Newport to Brynmawr: this runs under the Crumlin viaduct, and up the Ebbw fach; (iv) a branch of the former up the Ebbw fawr from Aberbig to Ebbw Vale; (v) the L. and N. W. R. line up the Sirhowy (Sorwy) from Newport to Tredegar and Nantybwch where it joins the Abergavenny and Merthyr; (vi) the Brecon and Merthyr company's line from Newport to Bargoed, Dowlais Top, and Brecon: this has a branch on the left bank of the Rhymney to New Tredegar and Rhymney. The Rhymney line from Cardiff to Rhymney Bridge, keeping as it does to the right bank of the river, is in Glamorganshire.

22. Administration and Divisions— Ancient and Modern.

In Wales the English Hundred and Township were represented by the Cantref and the Cwmmwd. Gwent was divided into two Cantrefi, Cantref Gwent including the southern part of the modern county, and Cantref Iscoed the northern part. The western hills formed Cantref Gwynllwg (Wenllwch)[1].

[1] Under feudal tenure the Cantref and Cwmmwd do not appear to have had any place in the administration, at any rate in Gwent.

Under Henry VIII the new county of Monmouth was divided into the hundreds of Abergavenny, Skenfrith, Raglan, Usk, Caldicot, and Gwynllwg (Wenllwch). Each hundred contains so many parishes. The hundred is now obsolete for administrative purposes, except for districts of coroners. In modern times a distinction has been made

Town Hall, and Statue of Henry V, Monmouth

between the civil parish and the ecclesiastical parish. The latter is the ancient division, and is used to signify the area attached to a particular church called the parish church. This was the ancient administrative unit. But in the last century, some of the old parishes had become so reduced in point of population (in some the very church has vanished) that they were united with other parishes to form the administrative unit called a civil parish, that is the area for which a separate poor-rate is, or can be levied, or for which a separate overseer can be appointed. Hence it follows that where the population of the old parish is large enough to come under this definition it will be identical with the civil parish, where it is not large enough it will only be a part of one, and where it is too large it will be in more than one. In our county the number of civil parishes is 159, of ecclesiastical, including Ecclesiastical Districts created by Order in Council, 135.

Monmouthshire is on the Oxford circuit, and the assizes are held at Monmouth, which, though far from having the largest population, still remains the county town.

The Quarter Sessions meet at Usk. This court consists of the magistrates of the county, but its functions since the establishment in 1888 of County Councils have been considerably reduced, and are now confined to criminal business and the licensing of public houses. They also managed the roads and bridges and controlled the lunatic asylums, but these duties now belong to the County Council, which meets at Newport. The county gaol is at Usk, and the police are controlled by a joint standing committee of the Quarter Sessions and County Council. The County Council consists of 68 members, of whom 17 are aldermen, who during their term of office (which is longer than that of a councillor) do not require re-election, thus giving an element of continuity to the assembly. Newport, having a population large enough to rank as a "county borough," is exempt from the jurisdiction of the County Council, and is governed by its own municipal authority. The so-called County Courts were tribunals instituted in 1846 for the settlement of small debts, and their districts are mapped out irrespective of county boundaries.

The local government of small areas is administered under the County Council by the Urban District (in the case of towns that are not boroughs), Rural District, and Parish Councils: the last has encroached on the duties of the Parish Vestry, a body comprising all the ratepayers in the parish, and therefore a strictly democratic institution; in practice however the attendance at its meetings is very small, and its proceedings do not go much beyond the election of churchwardens and the disposal of church funds.

There are five Poor Law Unions—Abergavenny, Bedwellty, Chepstow, Monmouth, and Pontypool, and of these Chepstow includes a few border parishes in Gloucestershire, and Monmouth a few in both Gloucestershire and Herefordshire.

The municipal boroughs are three—Monmouth, with a population of 5,269, Abergavenny 8,511, and Newport 83,700. These three places do not however combine to form "the Monmouth parliamentary district," which returns one member to the House of Commons, for in this association the place of Abergavenny is taken by Usk, the corporation of which was dissolved by the Municipal Corporations Act of 1883. The county, apart from this "district," returns three members, one for each of the three divisions—northern, western, and southern.   Geographically the southern division is the largest, while the western, which comprises the dense population of the northern portion of the mining districts, is the smallest.

The two chief officers of the county are the Lord Lieutenant (who is also the Custos Rotulorum[1]), and the Sheriff.   The one may be said to represent the king and the other the county.   Nevertheless both have for many centuries been appointed by the Crown.   The Lord Lieutenant is appointed for life : he is the head of the magistracy and recommends to the king persons fit for appointment as justices.   Formerly he had the control of the territorial forces, but, though he still recommends fit persons for commissions, this has for the last forty years been transferred to the Crown.   The Sheriff is the executive officer of the county ; he is chosen from the wealthy commoners every year " on the morrow of St Martin's," November 12th.   It is his duty to attend the judges at the assizes, and to see that their sentences are carried into execution.   He also presides over the election of members of Parliament, and in the case of a vacancy applies to the Speaker of the House of Commons for a new writ.

Monmouthshire is in the diocese of Llandaff, which also includes the greater part of the county of Glamorgan. From the eleventh century down to 1843, when it was restored to Llandaff, that part of the town of Monmouth

[1] *Keeper of the Rolls*, i.e. the records of the county sessions.   As a matter of fact they are kept by the Clerk of the Peace, an officer formerly appointed by the Custos, but now by the County Council.

which lies between the Monnow and the Wye was in
the diocese of Hereford, and till the same year Cwmyoy
and Llanthony were in that of St David's.   The whole
of the county is in the archdeaconry of Monmouth, which
is divided into ten rural deaneries.   The archdeacon has
a general jurisdiction under the bishop ; under him are
the rural deans, whose duty it is to call together at stated
times the clergy of their deaneries, and to see that the
churches and their furniture are kept in proper order.

23. Roll of Honour.

Our list of worthies may begin with the only king
of England ever born within our limits.   Henry V was
born at his grandfather's castle of Monmouth in 1387.
There was at this time no likelihood of his ever suc-
ceeding to the throne, his father only bearing the courtesy
title of Earl of Derby.   In 1399 he was created Prince
of Wales.   In 1403 he was appointed his father's lieutenant on the Marches of Wales for the purpose of
putting down the rising of Glyndwr.   He was not
present at the fight of Grosmont in March 1405, but
in the following May he defeated the Welsh in the
neighbourhood of Usk, slaying 1500 of them and taking
prisoner Glyndwr's son Gruffydd.   The rest of his career
is part of the history of England.

Sir William Herbert was the son of Sir William ap
Thomas of Raglan and, through his granddaughter, an-
cestor of the Dukes of Beaufort.   He was a staunch

supporter of Edward IV, by whom he was created Earl of Pembroke. He led 18,000 Welshmen to oppose the northern rebels in 1462, but was defeated at Danesmoor near Banbury, and together with his brother Sir Richard

Henry V

Herbert of Coldbrook conveyed by the rebels to Northampton and there beheaded: he lies buried at Tintern, his brother at Abergavenny. From his great-grandfather, Gwilim ap Jenkin, are descended all the race of Herbert.

Sir Roger Williams (died 1595) was a gallant soldier,

the son of Thomas Williams of Penrhôs Fwrdios near Caerleon. He served under the Earl of Leicester in the Netherlands, and distinguished himself by his defence of Sluys in 1587 against the Prince of Parma. It was in vain that the Prince tried to bribe him to enter his service; his reply was that his sword was first to serve her Majesty, and then Henry of Navarre.

The Herbert Chapel, St Mary's, Abergavenny

Henry Somerset (1577–1646), fifth Earl and first Marquess of Worcester, is famous as the chief supporter in the county of the Royal cause in the Civil War. He held his house at Raglan for the King throughout the war, and only surrendered it in August 1646, when, after a siege of ten weeks, it was impossible to defend it longer.

His son Edward Somerset (1601–1667), second Marquess, is best known as the author of *A century of the names and scantlings of such inventions as at present I can call to mind to have tried and perfected*, written in 1655 and printed in 1663. The sixty-eighth invention, *An admirable and most forcible way to drive up water by fire*, enunciated the principles of generating force by steam which afterwards developed into the steam-engine. In 1644, while still styled "Lord Herbert of Raglan," he had the title of Earl of Glamorgan conferred on him by the king and was sent to raise forces in Ireland, in which however he was unsuccessful. His successor in the marquisate was his son Henry Herbert (1629–1700), who came to terms with Cromwell, and became a member of his Council of State in 1651. After Cromwell's death he again courted the rising sun and for the rest of his life remained loyal to the House of Stuart. In 1672 he was appointed Lord President of the Council of Wales and the Marches, and in 1682 he was created Duke of Beaufort. He was the first of his family to own and reside at Badminton.

Sir Trevor Williams of Llangibby (died 1692) was a man well known in Monmouthshire during the Civil Wars. He was at first on the side of the Royalists, but we afterwards find him at the head of the local " club men " in arms for the Parliament. In January, 1646, he was appointed commander-in-chief of the Parliament's forces in Monmouthshire, but in 1648 he had gone round again and Cromwell was informed that he was " very deep " in the plot for betraying Chepstow Castle to Sir Nicholas

Kemeys. Cromwell considered him "a dangerous man, full of craft and subtlety; very bold and resolute"; and wrote from Pembroke a letter containing particular instructions for his arrest. He afterwards compounded as a Delinquent.

Sir Thomas Morgan (d. 1679), one of the bravest and most capable officers on the Parliament side in the Civil War, was the son of Lewis Morgan of Llangattock Lingoed. He served in the Thirty Years War, and in 1645 succeeded Massey as Governor of Gloucester, and was untiring in his efforts to bring the war to an end in that part of the country. He captured Chepstow and Monmouth from the Royalists, and in March 1646 was one of the victorious commanders at the battle of Stow on the Wold—the last action of the war fought in the open field. He then invested Raglan and directed the operations till the arrival of Fairfax in the last month of the siege. He was highly valued by Cromwell, but he sympathised with the Restoration, and was made Governor of Jersey by Charles II, in which office he effectually frustrated all the designs of the French upon that island.

His brother Sir Henry Morgan (1635 ?–1688) was a famous buccaneer in the West Indies, but his attacks on the Spaniards were authorised by commissions from the Governor and Council of Jamaica. In 1668 he sacked the town of Porto Bello on the north coast of the Isthmus of Panama, where it was believed that Spanish levies for an attack on Jamaica were being raised. In 1670 he defeated the Spaniards in a pitched battle and took the city of Panama. In the same year by the Treaty of Madrid Jamaica was formally ceded to England, and four years later Sir Henry was appointed Lieutenant-Governor of the island, senior member of the Council and com-

Field Marshal Lord Raglan, K.C.B.

mander-in-chief of the forces there. He died at Port Royal in 1688.

Fitzroy James Henry Somerset (1788–1855) was the youngest son of the fifth Duke of Beaufort and created Baron Raglan in 1852. He served with distinction under

Wellington in the Peninsula and at Waterloo. In the Crimean War he held the post of commander-in-chief till his death before Sebastopol. He was born at Badminton, but had a Monmouthshire home at Cefntilla near Usk, formerly the headquarters of Sir Thomas Fairfax during the siege of Raglan.

Sir Charles Hanbury Williams (1709–1759) of Coldbrook was the son of John Hanbury of Pontypool, and assumed the name of Williams under the will of his godfather, Charles Williams of Caerleon. He was a skilful diplomatist and represented the Court of St James's at Dresden, Berlin, and St Petersburg. He is also known as one of the wits of his day, and as the friend of Lord Holland, Horace Walpole, and George Selwyn. He wrote odes and satirical verses, which had a great vogue at the time, but are now forgotten.

William Blethyn (d. 1590) came of an old Welsh family and was born at Shirenewton. Archdeacon of Brecon and Prebendary of York he became Bishop of Llandaff in 1575. Another Monmouthshire bishop was William Bradshaw (1671–1732). He was born at Abergavenny, was educated at Balliol College, Oxford, and became Dean of Christ Church and Bishop of Bristol in 1724.

These are our only bishops; the other noted Divines of the county have for the most part been ranged on the Puritan side, but Lewis Evans (fl. 1574), first a Roman Catholic and then an Anglican controversialist, and Philip Evans (1645–1679), a Jesuit and one of the victims of Oates's plot, wei. natives of Monmouthshire.

Another victim of the plot was David Lewis (1617–1679) who was the son of the master of the Grammar School at Abergavenny. When he was about nineteen he became a Roman Catholic, and was afterwards a Jesuit missionary for 28 years among the persecuted Catholics in South Wales. At last he became head of a Jesuit Society at Llanrothal on the Herefordshire side of the Monnow, where he was arrested in 1679. He was then found guilty of high treason, under a statute of Elizabeth, at Monmouth Assizes, and hanged at Usk.

William Wroth (1576 ?—1642) was born near Abergavenny of good family and educated at Oxford. He became rector of Llanfaches and of Llanfihangel Roggiett, but he seems to have resided at the former place, and to have attracted such crowds thither by his eloquence that he was often compelled to preach in the churchyard. This and his Puritanical leanings brought upon him the censure of his bishop: he was summoned before the Court of High Commission and deprived of his living. Three years before his death he founded at Llanfaches the first Nonconformist community in Wales. The Congregational Chapel there still preserves his memory.

Walter Cradock (1606 ?–1659) was born and died at Trevella, a house on the northern slope of Golden Hill, in the parish of Llangwm Uchaf. He was educated at Oxford and took Holy Orders, but he sympathised with Wroth, and succeeded him as minister of the chapel at Llanfaches. In 1654 he was one of the Commissioners of Triers for the appointment of public preachers.

Edmund Jones (1702–1793), Yr Hen Prophwyd (the Old Prophet), was minister of the Ebenezer chapel at Aberystruth, of which parish he was a native. His father was named John Lewis, but according to the old Welsh custom, then fast becoming obsolete, the son took the father's Christian name as his surname. Jones was celebrated as a preacher, but he was also a writer, his principal works being *A relation of Ghosts and Apparitions, which commonly appear in the Principality of Wales*, 1767, and *A geographical historical and religious account of the parish of Aberystruth*, 1779.

Among the monuments in Abergavenny Church are those of two Monmouthshire lawyers, David Lewis (c. 1520–1584) and Andrew Powell (1565–1631). Lewis was the son of a vicar of Abergavenny, and became a Fellow of All Souls College, Oxford. When Jesus College was founded in 1571 he was appointed Principal. In 1554 he represented his native county in Parliament, and in 1558 was raised to the bench as a judge of the Admiralty Court. Powell, whose monument bears the effigies of himself and his wife, was born at Trostry near Usk, and became in 1607 one of the two judges of the Brecon circuit.

He was succeeded in this office by Walter Rumsey of Usk (1584–1660)—an astute lawyer, known as the *Picklock of the Law*. He was removed from his post by the Parliament in 1645, and died at Llanover. Sir Richard Morgan of Blackbrook in Skenfrith was member for the city of Gloucester in the Parliaments of Edward VI. He was a supporter of Queen Mary and was made Chief

Justice of the Common Pleas in 1553. In the same year he presided at the trial of Lady Jane Grey, and passed sentence on her. Two years later, says Foxe the martyrologist, he fell mad, and in his ravings cried out to have the Lady Jane taken away from him. He died in 1556.

Sir Thomas Phillips (1801–1867) was a solicitor at Newport, but was afterwards called to the bar. In 1839 he was Mayor of Newport and was wounded while reading the Riot Act, when John Frost and his 7000 Chartists attacked the Westgate Inn. He interested himself greatly in Welsh education and built schools for the colliers near Newport.

Geoffrey of Monmouth (1100?–1154) was probably educated at the Benedictine Priory in Monmouth, and became Archdeacon of Llandaff. His romance, which he called *Historia Britonum*, was compiled from Nennius and a lost book of Breton Legends. It was one of the great story-books of the Middle Ages, and has proved a storehouse of materials to a long series of poets. Milton was one of the first to impugn its authenticity. Two years before his death he was made Bishop of St Asaph, but he never visited his diocese.

If Geoffrey may be called the first of our local antiquaries, the list of his successors is a very short one. Not that plenty of books have not been written about the county, but with a few exceptions they have been written by strangers. Nathan Rogers (1639–c. 1710), of the parish of Llanfaches, was educated at St John's College, Oxford, and was by profession an attorney. In 1659

he sat in Parliament for the city of Hereford, and in
1708 he published *Memoirs of Monmouthshire*, a short
account of the history of the county drawn up in the
interests of the tenants of Wentwood who had suffered
from the encroachments and enclosures of the lords of
the manor, the first Duke of Beaufort and his predecessors.
Edmund Jones, an Independent minister of Aberystruth,
published in 1767 an account of the popular superstitions
of the Welsh, and in 1779 an historical account of the
parish in which he lived.   Charles Heath (1761–1831),
though a native of Worcestershire, established himself as
a printer in Monmouth and published several volumes on
the topography of that town and neighbourhood in which
he has preserved much information that would otherwise
have perished.   Charles Octavius Swinnerton Morgan
(1803–1888), M.P. for Monmouthshire 1841–1874, an
accomplished antiquary and the author of several mono-
graphs on the antiquities of his native county, was a
brother of the first Lord Tredegar.   In his local researches
he was assisted by Thomas Wakeman (1791–1868), also
a member of a Monmouthshire family.

    The only poet of any note that Monmouthshire has
produced is William Thomas (1832–1878)   From his
native place, Mynyddislwyn, where he worked as a
Calvinistic methodist minister, he took the bardic
name of Islwyn.   His Welsh poems are considered the
finest of the nineteenth century, the longest being a
philosophical poem entitled *Yr Ystorm (The Storm)*.   They
were collected in 1897 under the title of *Gweithiau
Islwyn*.

Lady Llanover (1802–1896) wife of Sir Benjamin Hall, Lord Llanover, and daughter of Benjamin Waddington of Llanover, was well known for her encouragement of the Welsh language, music, and customs.

Sir John Hopkins (1715–1796) of Llanfihangel-ystern-llewern may be mentioned as the only Lord Mayor of London whom the county ever produced.

24. THE CHIEF TOWNS AND VILLAGES OF MONMOUTHSHIRE.

N.B. The character of the mountainous region in the west of the county, known in the eighteenth century as "the Wilds of Monmouthshire," has during the last hundred years undergone a complete transformation, owing to the rise or the mining industries. New towns and villages have sprung up on all sides. The ancient parishes west of the Afon Lwyd and north of Risca are Bedwellty, Mynyddislwyn, Llanhiddel and Aberystruth.

(The figures in brackets after the name of the place give its population, while those at the end give the references in the text.

The full census figures for 1911 being as yet unpublished, the populations here given are those of 1901, except in the case of the larger towns, where the 1911 figures are available and are marked with an asterisk.)

Abercarn (*16,445), a large village containing collieries, iron, tinplate, and chemical works, formed into a civil parish under the Local Government Act of 1894 from Mynyddislwyn. (pp. 26, 65.)

Abergavenny (*8511), the largest of the ancient towns in the county after Newport, is situated at the confluence of the Gavenny with the Usk (Aber meaning "a confluence"). Since 1899 it has been a municipal borough. The church of St Mary, which has been partly rebuilt, was the church of the Benedictine Priory founded here in the reign of Henry I: it contains many

The Vale of Abergavenny from the East

interesting monuments. The parish of Holy Trinity was formed out of the mother parish in 1895. The castle, of which the ruins of two towers remain, was founded shortly after the Conquest by a Norman knight, Hammeline Baladun. The town contains breweries, corn mills, lime and stone works, iron foundries, and engine works. (pp. 13, 24, 25, 47, 48, 82, 92, 95, 105, 112, 114, 119, 120, 136, 142, 144, 146, 149, 154, 155.)

Abersychan (9436, *24,661), at the confluence of the Sychan with the Afon Lwyd, was cut out of the parish of Trevethin in 1844. It is two miles north-west of Pontypool. (p. 25.)

Abertillery (16,930, *35,425), a town and civil parish cut out of Aberystruth under the Act of 1894, 18 miles north of Newport, and containing tinplate works and collieries. Cwmtillery is a hamlet with a colliery population a mile and a half to the north. (p. 26.)

Aberystruth, otherwise called Blaenau Gwent (13,491), an ancient parish nine miles west from Abergavenny and containing Nant-y-glo, Garnfach, and Blaenau. The present church was built in 1857 on the site of the ancient church burnt down a few years previously. This is described by Coxe in 1800 as a handsome building with a square tower, "peculiarly striking from its sequestered situation [in the midst of fields] and singular appearance; the outside of the body and chancel, with the lower part of the tower and its battlements, are whitened; the remaining part of the tower is of hewn stone uncoloured." It is hardly necessary to add that the "fields" have now given place to iron furnaces and collieries. Nant-y-glo (4704) is two miles north and was formed into an ecclesiastical parish in 1844. The population work in the collieries. (pp. 156, 158.)

Bassaleg (4240), a parish on the Ebwy (Ebbw) about three miles west of Newport, with four railway stations. On the other side of the Ebwy is the hamlet of Rogerstone. (pp. 82, 121.)

Bedwas (civil parish 2080, ecclesiastical 1855), on the Rhymney, two miles north of Caerphilly. The hamlet of Ruddry and 571 of the inhabitants are on the Glamorganshire side of the river. (pp. 13, 27.)

Bedwellty (*22,551), a large parish containing the villages of Aberbargoed, Argoed, and Blackwood, 16 miles north-west of Newport, and the seat of many iron-works and collieries. The church stands on the top of the ridge (1000 ft.) which separates the valleys of the Rhymney and the Sorwy (Sirhowy). When Coxe saw it more than a century ago he found an "embattled tower built with brown rubble [ironstone?], and coigned with hewn stone; the battlements as well as the body whitewashed." Though only reached by a stiff climb it is still the favourite church with the colliers for marriages. (pp. 27, 130, 146.)

Bettws-Newydd (75), a small parish on the road from Usk to Abergavenny: the church retains a beautiful rood screen. (p. 111.)

Blaenavon (*12,010), at the head of the Afon Lwyd, the seat of the extensive coal, iron, and steel works of a company which takes its name from the place. There are nine blast furnaces, three rolling mills, and one tyre mill, which employ about 5000 persons. The ecclesiastical parish was formed in 1860 out of Llanfoist, Llanwenarth, and Trevethin. The church was built in 1804 by Messrs Hopkins and Hill, who at that time owned the works. (pp. 25, 143.)

Blaenau, see **Aberystruth**.

Caerleon (*2046), the Isca Silurum of the Romans, lies on the Usk 2½ miles above Newport. A portion of the Roman wall may still be seen, and outside it is the amphitheatre, in which excavations have been commenced. In the Museum is a collection of Roman antiquities. The mound of the medieval castle has unfortunately been planted with trees. The ancient church

Caerleon

of Llangattock-juxta-Caerleon (230) has been mainly rebuilt.
(pp. 24, 25, 70, 78, 79, 92, 93–95, 105, 111, 113, 114, 136, 150,
154.)

Caerwent (369), the Venta Silurum of the Romans, is now
a small village. The Roman walls are still fairly preserved on
all four sides of the former town, and considerable excavations
have taken place in the last few years, which have enabled the
plan of the ancient town to be traced. (pp. 79, 92, 93–95.)

Caldicot (1196), a considerable village rather more than a
mile from the Severn Tunnel Junction, has a fine church with
a central tower, but no transepts. The castle, of which the shell
remains, dates from the twelfth century. The gatehouse has
been restored and is inhabited. Near the Channel are some
extensive tinplate works. (pp. 33, 39, 96, 107, 109, 111, 119,
144.)

Chapel Hill (309) deserves mention as the parish which
contains the ruins of Tintern Abbey, founded 1131. (pp. 82,
95, 97, 102.)

Chepstow (*2953) is built on the side of a hill sloping down
to the Wye, less than three miles from its mouth. It formerly
had a considerable trade by water with Bristol and other places.
There is now an iron foundry here, which employs a large
number of people. The salmon fishery has always been a famous
one. Formerly it belonged to the Duke of Beaufort as lord of
the manor, but it is now Crown property. The chief attractions
of the town are the fine Norman church, once the church of a
Benedictine Priory, of which the nave and west front remain,
and the extensive ruins of the castle founded soon after the
Conquest. The port wall which surrounds the town (the other
side being defended by the river) is still tolerably perfect, and
includes a gatehouse through which the town is entered from
the south-west. Brunel's great tubular railway bridge is a blot

Tintern Village

upon the landscape; on the other hand no more elegant iron bridge exists than that by which the coach road crosses the river. (pp. 23, 33, 51, 61, 65, 67, 70, 71, 81, 84, 85, 91, 95, 99, 103, 113, 114, 115–119, 123, 135, 136–142, 146, 152.)

Christchurch (1296), on the top of the hill three miles north-east of Newport, with a large church. The greater part of the parish, including the ecclesiastical parish of Maendu (Maindee) formed in 1860 and now in the civil parish of Newport, is comprised in the borough of Newport. (pp. 12, 16, 96, 107, 110.)

Crumlin, see **Mynyddislwyn**.

Cwmbran, see **Llanfihangel Lantarnam**.

Cymyoy (332), a parish on the Honddu nearly eight miles long and one wide; it contains the ruins of Llanthony Priory founded in the reign of Henry I.

Dixton (123), a parish on both sides of the Wye near Monmouth. It contains the famous Kymin hill (840 feet) from which ten counties are said to be visible. (p. 8.)

Ebbw Vale (*30,559), a large parish cut out of Aberystruth and Bedwellty, situated on the Ebbw Fawr, and 20 miles north-west of Newport. At the head of the valley are the works and collieries of the Ebbw Vale Steel, Iron, and Coal Company. Their output of pig-iron amounts annually to 250,000 tons, and of steel rails, etc. to 160,000 tons. The population continues to increase, and since 1895 upwards of 1000 new houses have been built. (pp. 65, 142, 143.)

Fleur-de-lis (2495), a modern parish with a fanciful name, taken from the sign of a public house on the Rhymney, formed in 1896 out of Mynyddislwyn, Bedwellty, and Bedwas. A small part on the right bank of the river was taken from the Glamorganshire parish of Gelligaer. Most of the inhabitants work in the neighbouring collieries. (p. 27.)

Chepstow

Goldcliff (253) a parish on the Bristol Channel containing the remains of a Benedictine Priory. (pp. 7, 33, 39, 40, 41.)

Griffithstown (3906), one mile south of Pontypool, was formed in 1898 out of the parishes of Llanfrechfa and Panteg.

Grosmont (523), on the Monnow. The church is a fine cruciform building with central tower and spire dating from the thirteenth century. Before the place was destroyed by the

Llanelen

partisans of Glyndwr it was probably of greater importance. The ruins of the castle still exist. (pp. 7, 30, 83, 96, 107, 108, 113, 120.)

Llandenny (322), on the Olwy, contains the ancient mansion of Treworgan, and also Cefntilla—the headquarters of Fairfax during the siege of Raglan.

Llandogo (487), on the Wye, including the hamlet of Whitebrook, where there were formerly paper-mills. The Cleiddon Shoots, a small waterfall, descend from the top of the hill (700 ft.) above the village. (p. 21.)

Llanelen (277) a small village south of Abergavenny with a fine stone bridge over the Usk.

Llanfair Discoed (149), a small village at the foot of the Grey Hill, two miles north-west of Caerwent, with ruins of a thirteenth century castle. (pp. 113, 114.)

Llanfihangel Lantarnam (5287), a parish four miles north of Newport. Here was a Cistercian abbey founded in the reign of Henry II, of which little but a gateway now remains. Cwmbran, a large village with iron-works and collieries, is in this parish. (p. 26.)

Llanfrechfa (4262), a large parish on both sides of the Afon Lwyd. The part on the west side of the river has been formed into a separate parish under the name of Llanfrechfa Upper, and includes the village of Pontnewydd. (p. 30.)

Llangibby (428), on the Usk, about three miles south of the town of Usk has an interesting unspoilt church, and the ruins of a castle. (pp. 111, 131.)

Llangwm Uchaf and **Isaf** (115), two parishes four miles east from Usk, now united. At Llangwm Uchaf is an interesting church with a beautifully carved rood screen of medieval date. (pp. 110, 111, 155.)

Llanhiddel or **Llanhilleth** (5015), a large parish five miles west of Pontypool, and near the junction of the Ebbw Fach with the main stream. Hard by are some ancient mounds and entrenchments known as Castell Taliorum. Part of Crumlin is in this parish. See **Mynyddislwyn.**

Llanmartin (158), contains the castle of Pencoed—a building which has more the character of a moated house than a castle. There is no trace of any work older than the thirteenth century, and most of it is of the fifteenth and sixteenth centuries. Much of the building is in ruins. (p. 114.)

Llantilio Crossenny (570), on the Trothy, eight miles above Monmouth. The church may rank with Grosmont and

The Brecon and Newport Canal, Llanwenarth

Trellech as one of the finest in the county. On a hill in the north of the parish are the ruins of White Castle. Not far from the church is the site of a moated house called Hen Gwrt, said to have been the residence of Sir David Gam, who fell at Agincourt. (pp. 96, 107, 108, 130.)

Llantilio Pertholey (1170), about two miles north of Abergavenny, with an interesting Perpendicular church.

Llanwenarth (1098), on the Usk above Abergavenny. In 1865 the part of the parish on the right bank of the Usk containing the village of Govilon was cut off from the mother parish under the name of Llanwenarth Ultra. In the latter parish is the hamlet of Pwl-du, inhabited by miners, colliers, and quarrymen in the employment of the Blaenafon Iron Company.

Grammar School, Monmouth

Machen (3367), a parish on both sides of the Rhymney (and therefore partly in Glamorganshire) eight miles from Newport, containing collieries, railway works, and a flannel manufactory. (pp. 27, 130.)

Magor (702), a village eight miles east of Newport, with a fine church. (pp. 32, 33, 96, 107, 110.)

Malpas (495) a parish north of Newport. There was a Cluniac Priory here founded in the twelfth century. The small Norman church was rebuilt in the last century, but its most interesting features—the south doorway and chancel arch—were preserved. (p. 111.)

Marshfield (581) a parish on the road between Newport and Cardiff, with a fine Norman and Early English church.

Mathern (502), about two miles south of Chepstow. The church is dedicated to St Tewdric, a canonised king of Glamorgan mortally wounded in a battle with the Saxons at Tintern in the sixth century, who died here. It is a fine Early English building with Perpendicular additions, including a tower of exceptional dignity in this district. The former palace of the Bishops of Llandaff, and an ancient mansion of great beauty, Moynes Court, are in this parish not far from the church. (pp. 96, 107, 109, 128.)

Michaelstone-y-fedw (506), a parish on both sides of the Rhymney and therefore partly in Glamorganshire—a circumstance which shows how the parish boundaries on this side were disregarded when the county of Monmouth was created. The transept of the church is the ancient burial-place of the Kemeys family of Cefn Mabli on the other side of the river.

Monmouth (*5269), the county town, is situated at the confluence of the Monnow (Mynwy) with the Wye. Both rivers are crossed by stone bridges, that over the Wye bears the date 1617, and that over the Monnow is still older and has a fortified gatehouse of two storeys. The church, with the exception of the tower and spire, was rebuilt in 1883, the original Early English church having in 1737 been replaced by an ugly building in the style of that time. Over Monnow, with its modernised Norman church of St Thomas the Martyr, was made into a separate parish in 1832. Of the castle only a few fragments remain, but a handsome house of the Restoration period stands within its precincts. The flourishing Grammar School (rebuilt 1864) was founded by William Jones of Newland in the Forest of Dean, a member of the Haberdashers' Company, in 1614. The Girls'

High School, which is managed by the Governors of the Grammar School and four lady Governors, was built in 1900. (pp. 4, 57, 82, 84, 92, 95, 99, 107, 114, 115, 119, 120, 135, 145, 146, 147, 148, 152, 155, 157.)

Mynyddislwyn (*9982), a large parish 10 miles north-west of Newport. The church, which was rebuilt in 1820, is situated on the hills about halfway between the Ebwy and the

The Garn Lwyd, Newchurch

Sorwy; near it is a tumulus. The part of Crumlin which lies west of the Ebwy is in this parish 12 miles north of Newport. The Crumlin viaduct, built about 1857, carries the Pontypool and Aberdare railway across the gorge of the Ebwy, at a height of 200 feet: the length of the iron-work is 1500 feet, or including the stone abutments, 1658 feet. See also **Abercarn** and **Fleur-de-lis**. (pp. 26, 27, 143.)

Newport

Nant-y-glo, see **Aberystruth.**

Newchurch (817), a very large parish of 5497 acres lying on the lofty ridge which bounds the basin of the Usk on the south. The dolmen called the Gaer or Garn Lwyd, the ancient camp called the Gaerfawr, and Troggy castle, an outpost built in the thirteenth century by the lord of Striguil, are in this parish. In the east of the parish is the Devauden (Ddfawedun), a hamlet which was the scene of the labours of the locally famous schoolmaster James Davies. (pp. 16, 24, 90, 91.)

Newport (67,270, *83,700), constituted in 1891 a county borough, lies on the Usk about four miles from its mouth. There was no doubt a castle here in the twelfth century, but the building of which the mutilated river frontage, 228 yards in length, remains does not appear to be older than the fourteenth century. There are now, since 1836, several churches. The mother church of St Woollos on Stow Hill is a large edifice partly of Norman date, consisting of a nave, aisles, and chancel connected with the western tower by a kind of antechapel 43 feet long. There are, apart from the railway, two bridges, one the stone bridge built in 1800 in place of the former wooden bridge, the other a "transporter" bridge from which a travelling car is suspended. The extensive docks are still being enlarged. In 1839 Newport was the scene of a Chartist riot under the leadership of one John Frost, which was put down by the aid of the military. Newport now ranks with Cardiff and Swansea as one of the great trading ports of South Wales. (pp. 5, 14, 24, 26, 48, 59, 61, 63, 65, 67, 68, 70, 71, 73, 75-78, 85, 96, 103, 114, 115, 119, 123, 143, 145, 157.)

Panteg (*10,099), between Newport and Pontypool, containing coal mines, stone quarries, and foundries. The canal is here carried over the Afon Lwyd by an aqueduct.

Penmaen (5846), an ecclesiastical parish 18 miles northwest of Newport, and formed out of Mynyddislwyn in 1845.

Pontnewynydd (4168), on the Afon Lwyd, formed into an ecclesiastical parish in 1845 out of Trevethin. The village contains coal mines and a large steel and galvanized iron manufactory.

Pontypool (*6452), one of the principal centres of the coal and iron trade, lies on the slope above the Afon Lwyd. Ecclesiastically in the parish of Trevethin on the opposite side of the river, it is now a civil parish of itself. The town owes its origin to the ironworks planted here towards the end of the sixteenth century by the Hanbury family, which settled at a later date at what has since been known as Pontypool Park on the other side of the river. In the neighbourhood are numerous forges and iron-mills, but the ore now used is imported from abroad, and chiefly from Spain. See Trevethin. (pp. 26, 58, 59, 62, 131, 135, 142, 143, 146, 154.)

Portskewet (868), a village five miles south-west of Chepstow, with an interesting Norman and Early English church. The parish of Sudbrook with its ruined church and camp is now united with Portskewet. The western end of the Severn tunnel reaches the shore in this parish, but its mouth is in Caldicot. This tunnel was begun in 1879 and finished in 1886; its total length is 4 miles 634 yards, of which $2\frac{1}{2}$ miles are under the bed of the Channel. The pumping works at Sudbrook exhaust the freshwater springs which drain into the tunnel and raise from 20 to 25 millions of gallons a day. The fan for ventilating the tunnel is 40 feet in circumference. (pp. 39, 90, 91, 97.)

Raglan (619), in the centre of a large depression surrounded on all sides by hills, and famous for the ruined mansion of the Marquess of Worcester, which sustained a siege at the end of the first Civil War. In the church are mutilated monuments of the Somerset family. (pp. 13, 84, 123, 124, 144, 150.)

E. M. 12

Rhymney (*11,451), a parish cut out of Bedwellty in 1843. It is situated on the upper Rhymney, and its only production is coal. (p. 143.)

Risca (*14,149), on the western branch of the Monmouthshire canal about seven miles north-west of Newport, containing collieries and manufactories as well as many acres of tilth and pasture. The Urban District, which is divided into central, north and south wards, was formed in 1878 out of the parishes of Risca, Machen, and Mynyddyslwyn. (pp. 26, 65.)

Roggiett (118), near the Severn Tunnel Junction. The church is large and has a fine western tower with an octagonal pinnacle in one corner. (p. 110.)

St Arvans (536), a village lying between the hills and the Wye two miles above Chepstow and containing the well-known Piercefield Park and woods laid out by Valentine Morris in the eighteenth century.

St Mellons (637), a parish in the south-west of the county with an interesting church, chiefly of the Decorated period.

Shirenewton (779), a large parish on the top of the hill to the east of the Mynydd Lwyd (Grey Hill), with an interesting church of the thirteenth century. (pp. 110, 154.)

Skenfrith (426), on the Monnow. The church is fine and unspoilt. The ruins of the castle comprise a central keep built on a mound and a strongly fortified bailey. (pp. 10, 107, 109, 110, 113, 114, 120, 130, 144, 156.)

Tintern-Parva (288), on the Wye above Tintern Abbey. In the valley of the Angidy which joins the Wye here were formerly wire-works and an iron forge. The iron railway bridge connected with the former undertaking is still an eyesore in the landscape. The cottages of this village and of the adjoining parish of Llandogo are dotted about the steep woody banks of the river to their summit. For **Tintern Abbey,** see **Chapel Hill.** (pp. 58, 70, 149.)

Tredegar (*23,604), a large modern town (not to be con-fused with Tredegar Park) which sprang up at the beginning of the nineteenth century in the head of the Sorwy (Sirhowy) valley owing to the establishment here of the ironworks belonging to the Tredegar Company. The ecclesiastical parish was cut out of Bedwellty in 1840. The principal streets diverge from "the Circle" in the centre of the town. The inhabitants are colliers and ironworkers. New Tredegar (5797) on the Rhymney was cut out of Tredegar in 1890. (pp. 27, 33, 65, 90, 131, 143.)

Trellech (682), a village standing on the range of hills above the Wye below Monmouth. From its large church it may be inferred that it was once a town of some importance. The nave is divided from the aisles by a fine arcade of five bays supporting a clerestory, and the western tower is crowned by a lofty spire. Within the village is one of those mounds which are believed to have been watching mounds, and just outside the village are three menhirs or hoar stones. The original place of this name was at what is now called Trellech Grange (Tre-llech, the vill of the flat stone) and when the present town was founded in the twelfth or early thirteenth century the name was transferred to it. (pp. 12, 16, 32, 54, 59, 90, 96, 107, 108, 111.)

Trevethin (including Pontypool, 10,380). This is the mother parish of Pontypool. The church, on the hill across the Afon Lwyd from the town, was rebuilt in 1846, with the exception of the tower.

Usk (*1495), a small town in the centre of the county. The river Usk is here crossed by a fine stone bridge, and the view hence up the river to the mountains round Abergavenny was formerly very beautiful, but it has been spoilt by the intrusion of an iron railway bridge in the foreground. The church, of which the nave and central tower remain, was that of a Priory of Benedictine nuns: it contains a portion of a Welsh inscribed brass commemorating Adam of Usk, the fourteenth century chronicler. The ruins of the castle are in course of destruction by ivy. (pp. 13, 24, 25, 35, 59, 82, 92, 95, 107, 110, 114, 123, 144, 145, 154, 155, 156.)

Wolvesnewton (155), a parish in the hollow to the north of the Newchurch range, and east of Llangwm, so called as a new manor acquired by the family of Lupus or Lovell in the thirteenth century. It contains a moated stronghold called Cwrt-y-gaer. (p. 111.)

Wonastow (129), originally St Guingaloius or St Wennol, on the Trothy two miles south-west of Monmouth. It contains the interesting houses of Wonastow Court and Tre-Owen. (p. 109.)

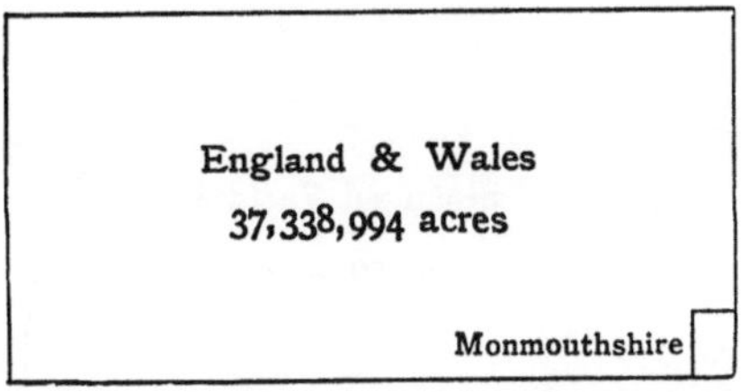

England & Wales
37,338,994 acres

Monmouthshire

Fig. 1.   The Area of Monmouthshire, 349,552 acres, compared
with that of England and Wales (1909)

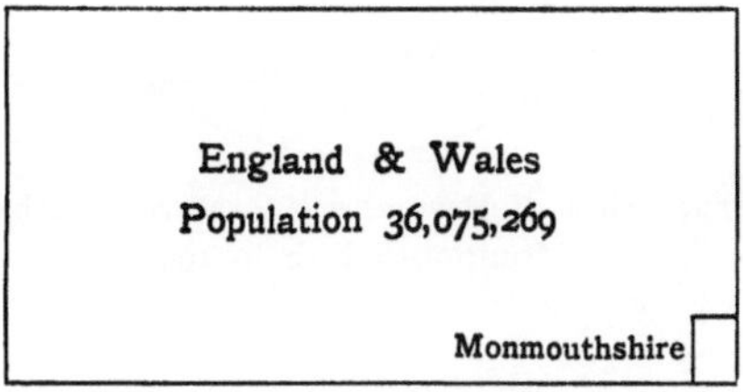

England & Wales
Population 36,075,269

Monmouthshire

Fig. 2.   The Population of Monmouthshire (395,778) compared
with that of England and Wales in 1911

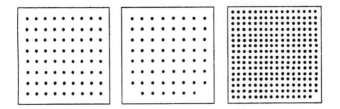

Monmouthshire, 724   England and Wales, 618   Lancashire, 2550

Fig. 3.   Comparative Density of the Population to the square
mile in 1911

(*Each dot represents* 10 *persons*)

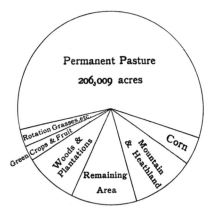

Fig. 4.   Proportion of Permanent Pasture to other areas in
Monmouthshire in 1909

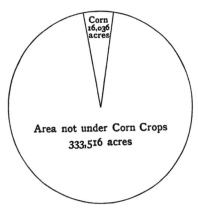

Fig. 5.   Proportionate Area under Corn Crops in
Monmouthshire in 1909

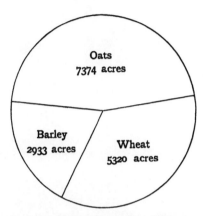

Fig 6. Proportionate Areas of Oats, Wheat, and Barley
in Monmouthshire in 1909

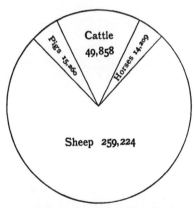

Fig. 7. Proportionate numbers of Sheep, Cattle, Horses,
and Pigs in Monmouthshire in 1909